Ridership Revisited:
The Official Ridership Forecast for the Proposed Baltimore-Washington Maglev Is a Factor of Ten Too High

Owen A. Kelley
Greenbelt, Maryland

Copyright © 2021 by Owen A. Kelley

Written and published by Owen A. Kelley in Greenbelt, Maryland, USA. The author may be contacted at okelley@gmu.edu. This document is based on five articles that were published in the Greenbelt Online blog in, 2021, https://www.greenbeltonline.org/maglev-ridership/. The cover photos were taken by the author except for the traffic photo (credit: Dave Dildine/WTOP) and the power-plant photo (credit: Adrian Jones, Integration and Application Network, https://ian.umces.edu/media-library/chalk-point-power-plant/). Back cover quotes: City of Greenbelt, 24 May 2021: *Comments on the Baltimore-Washington Superconducting Maglev Project Draft Environmental Impact Statement*, 222 pp, https://www.greenbeltmd.gov/maglev. Randal O'Toole, 6 April 2021: Maglev to destroy habitat, climate, *CATO At Liberty* blog, Cato Institute, https://www.cato.org/blog/meglev-destroy-habitat-climate. Disclaimer: This analysis was performed by an area resident, acting in his capacity as a individual citizen to examine a non-partisan issue of interest to the public. If any errors are found, please contact the author at okelley@gmu.edu. **ISBN:** 978-0-9670633-4-8.

maglev

(măg′lĕv, pronounced like *mag*netic *lev*itation) *n.* A kind of train that floats a few inches above its track and is propelled by powerful magnets.

maglev vs. SCMAGLEV

The exact kind of magnetic-levitation technology proposed for the Baltimore-Washington project is largely irrelevant when forecasting ridership. To a good approximation, two factors would determine the number of riders for any maglev line between downtown Washington, DC, and Baltimore. These factors are the duration of the maglev trip and the speed of travel on this region's existing roads and rail lines. For this reason, the present document refers to this rail proposal as the Baltimore-Washington maglev. Official documents refer to the project as the SCMAGLEV, which stands for "superconducting maglev." For the past sixty years, Japan has been developing superconducting-maglev technology. So far, no superconducting-maglev line has been built connecting two cities.

■ CONTENTS ■

Introduction 4

Part One • The Ridership Forecast and Supporting Analysis

    1 • Ridership 9

        The Federal Railroad Administration falls for an excessively high forecast of how many trips would be made on the proposed Baltimore-Washington maglev. Reference data suggest that the official ridership forecast is more than a factor of ten too high.

    2 • Wealth 27

        Most maglev riders would belong to the wealthiest 2% of the Baltimore-Washington population. An individual would earn at least $363,000 a year before finding the maglev ticket price worth the modest reduction in door-to-door travel time.

    3 • Geographic Area. 41

        The proposed Baltimore-Washington maglev would serve a small geographic area. Most maglev customers would start and end their trips near a maglev station, but the project's draft environmental impact statement is vague on this point.

Part Two • Maglev Impacts that Vary with the Ridership Forecast

    4 • Greenhouse Gas 69

        Operating the proposed Baltimore-Washington maglev would increase greenhouse gas emissions, the Federal Railroad Administration finds. According to the project's draft environmental impact statement, operating the maglev would increase annual carbon dioxide emissions by more than a hundred million kilograms, contradicting the claims of maglev promoters.

    5 • Road Congestion 83

        Data from the Federal Railroad Administration shows that building the proposed maglev would do little to reduce regional road congestion.

About the Author 91

■ INTRODUCTION ■

A federal regulator is currently considering whether to approve the construction of a rail line that a private company wants to build between Baltimore and Washington, DC. The train would use superconducting magnetic-levitation technology, hence its name: the maglev.

The federal regulator seems not to have noticed that the maglev proposal is fatally flawed because the project would attract insufficient ridership to justify its construction. The federal regulator is the Federal Railroad Administration, and the company that wants to build the maglev is Baltimore-Washington Rapid Rail (BWRR).

Public concern about the maglev's doubtful prospects was not resolved by the Federal Railroad Administration's six-page discussion of ridership in the draft environmental impact statement published in January 2021. The statement copied its ridership numbers from a report that the public is not allowed to read in full, a report that was written by a consulting company called Louis Berger.[1]

The present document breaks down this line of reasoning into five chapters, each of which makes one of the following points:

1. The proposed Baltimore-Washington maglev's official ridership forecast is more than a factor of ten higher than reference data can support.

The official estimate is 25 million one-way maglev trips per year, but reference data suggest that under 1 million maglev trips per year is more likely.[2]

2. Maglev riders would be predominantly wealthy, not a cross section of society.

Given the maglev ticket price and limited travel-time savings relative to car travel, only the wealthiest 2% of the region's population would likely ride the maglev.

3. The maglev would serve a small geographic area, not the entire Baltimore-Washington region.

The travel-time savings that matter are based on total travel time, door to door. Total travel time includes both the time to travel from the trip origin to a maglev station and from the

---

1. Public concern about the official ridership forecast is summarized on page 15 of the present document. The six-page description of ridership-forecasting methodology is found in DEIS Appendix D2, pages B-104 to D-109. The Louis Berger report is discussed on page 10 of the present document.
2. For 2045, the Federal Railroad Administration (FRA) forecasts 24.938 million one-way maglev trips per year if the Baltimore maglev station were built at Camden Yards: draft environmental impact statement (DEIS), Chapter 4.2, Table 4.2-3, page 4.2-7. Of this total, the DEIS forecasts that 20.6 million maglev trips would be diverted from other modes of transportation. The present document estimate that 1 million diverted trips is a more likely forecast, as stated in Chapter 1 (page 13).

final maglev station to the actual destination. The maglev would save travelers a significant amount of time within only a small area near each of the three maglev stations. The stations would be located in downtown Washington, BWI airport, and downtown Baltimore. For this reason, most counties in the Baltimore-Washington region would have few if any maglev customers start or end their trips there.

4. Both construction and operation of the maglev would increase greenhouse gas emissions and thereby thwart efforts to combat climate change.

Maglev operation would take so few cars off the road that greenhouse gas emissions would increase. The source of the maglev's greenhouse emissions is the generation of electricity to run the maglev. The draft environmental impact statement says as much, but the information is buried in an appendix. Furthermore, the statement does not even attempt to estimate greenhouse emissions from constructing the maglev. The present document fills this gap.

5. The maglev would do very little to reduce regional road congestion.

Even if the maglev's official ridership forecast were accurate, the amount of car travel that would be avoided once the maglev starts operating would be small. After less than a year, the natural, gradual increase in regional road traffic would erase the forecasted road-traffic reduction from maglev operation. These statistics on maglev travel and regional road traffic were published in the maglev's draft environmental impact statement, but the statement failed to put these two statistics together and draw the logical conclusion.

Consider the evidence for yourself as you read the ridership analysis in this document. All the relevant data are provided, and only simple arithmetic is used to compare and combine datasets. Each chapter's appendices provide mathematical details for planning professionals and others who want them.

RIDERSHIP
PAGE 9

WEALTH
PAGE 27

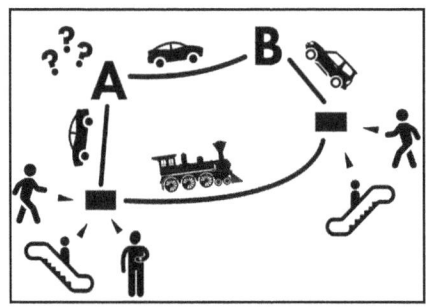

GEOGRAPHIC AREA
PAGE 41

PART ONE

The Ridership Forecast and Supporting Analysis

There is local criticism that the project was showy and wasteful, delivering no practical benefit to residents. Maglev ridership has been below expectations, due to limited operating hours, the short line, the high price of the tickets and the inconvenient location of the Longyang Road Terminal.

> —International Maglev Board (2007), describing China's Shanghai maglev, one of the few maglevs in commercial service in the world

1 • RIDERSHIP

The Federal Railroad Administration falls for an excessively high forecast of how many trips would be made on the maglev

Reference data suggest that the official ridership estimate is more than a factor of ten too high for the proposed Baltimore-Washington maglev

It would be a scandal to spend 17 billion dollars to build a new rail line if one could predict that the train would run mostly empty. There are hints that this disaster might unfold if a superconducting magnetic-levitation rail line were built between Baltimore and Washington, DC. Most people call this train the "maglev."[1]

The official ridership forecast for the Baltimore-Washington maglev is stated in the project's draft environmental impact statement. The statement, however, merely copies its ridership forecast from a contractor's report, a report that the public is not allowed to read. This secrecy makes it difficult, but not impossible, to double-check the official ridership forecast. If an approximate answer is sufficient, then only a few mathematical steps are needed to derive a ridership forecast consistent with the maglev's characteristics.

The accuracy of the official ridership forecast matters because the maglev's draft environmental impact statement relies on this forecast to quantify the various benefits of operating the maglev. The number of people riding the maglev determines the revenue from ticket sales, the financial solvency of the maglev operator, the amount of road congestion prevented, the reduction in car-generated air pollution, and the number of jobs created.[2]

It is unclear how low ridership would have to be to make the maglev worthless. The draft

---

1. $15–17 billion: FRA (2021), SCMAGLEV DEIS, Appendix D4, Table D4-8, pg. D-21.
2. Revenue from maglev ticket sales in "SCMAGLEV annual fare cost" row of Appendix D4, Table D4-28, pg. D-44. Road congestion: see page 83 of the present document. Air pollution: Appendix D4, Table D4-40, pg. D-51, and see page 69 of the present document. 390–440 jobs created by maglev operations: Chapter 4.6, pg. 4.6-8.

environmental impact statement ignores this question. Would this threshold be crossed if the official ridership forecast were, say, twice as high as would be reasonable? The analysis below suggests the official ridership forecast is more than ten times greater than can be supported by several datasets that describe the region's travel patterns.[3]

Background

In January 2021, the Federal Railroad Administration published the maglev's draft environmental impact statement. The document describes ridership as a "key metric" for determining the impacts of operating the proposed maglev. Curiously, the draft environmental impact statement uses only six of its 3,000 pages to describe its ridership forecasting method. Such a brief discussion of such an important topic is odd. The draft environmental impact statement provides so little detail that the official ridership forecast is not reproducible.[4]

Worse yet, it appears that the Federal Railroad Administration merely copied its ridership numbers from a contractor's report. The contracting company is named Louis Berger.

In the draft environmental impact statement, there is no evidence that the Federal Railroad Administration commissioned an independent review of the Louis Berger ridership report or had its staff perform an internal review. The draft environmental impact statement does mention one review of the Louis Berger ridership report, but that review suffers from a conflict of interest. That review was conducted by the company that wants to build the maglev, i.e., Baltimore Washington Rapid Rail (BWRR).[5]

By republishing Louis Berger's numbers in the draft environmental impact statement, the Federal Railroad Administration has transformed these numbers into the project's official ridership forecast.

The Louis Berger ridership report was completely hidden from the public during most of the public-comment period for the maglev's draft environmental impact statement. The company that wants to build the maglev, BWRR, was allowed to see the report but the public and elected officials were not. Toward the end of the public comment period, the Federal Railroad Administration made public a heavily redacted copy of the Louis Berger report. The information

---

3. Many ridership forecasts off ±30%: Hartgen 2013. A factor of 10 error would be unusually large.

4. A six-page-long ridership-model description citing zero references: Appendix D2, pg. B-104 to D-109. Key metric: Chapter 4.2, pg. 4.2-6. 654 pages in main text and 2,399 pages in the appendices, so the total page count is 3,053. To count pages, use the mdls command in the MacOS terminal: mdls -n kMDItemNumberOfPages *.pdf | awk '{print $3; sum += $3} END {print sum}'.

5. The maglev DEIS cites the 2018 Louis Berger "Baltimore-Washington SCMAGLEV Project Final Ridership Report" in Appendix D4 (footnote to Table D4-19, pg. D-36) and in Chapter 4.6 (pg. 4.6-3, footnote 9). The DEIS describes three steps that the "project sponsor" took to check the ridership forecast (Appendix D2, pg. B-104), but no steps that the Federal Railroad Administration took. The Federal Railroad Administration is a regulatory agency, so one of its essential functions is to double-check statements made by project sponsors, i.e., by the industry that the agency is supposed to be regulating. From the page following the title page of the draft environmental impact statement: "The Project Sponsor, Baltimore Washington Rapid Rail, LLC proposes to construct and operate an SCMAGLEV system between Baltimore, MD and Washington, D.C." See the discussion in Voulgaris (2019) on how a forecast can be affected by the biases of the forecaster.

relevant to the present chapter, however, was blanked out in the redacted version.[6]

The maglev's environmental impact cannot be estimated without first forecasting maglev ridership and diverted car travel. For this reason, the implausible and unsubstantiated ridership forecast in the maglev's draft environmental impact statement threatens to invalidate the entire impact statement. On this topic, an expert on impact statements said the following:

> The EIS [environmental impact statement] must be written in a manner that can be readily understood by the decision maker and the public. Yet, at the same time, it must provide an "accurate," "rigorous," and "scientific" analysis of environmental impacts (Sections 1500.1[b] and 1502.14[a]). Failure to comply with either of these opposing goals may provide a basis for successful litigation.[7]

The above quote refers to Parts 1500 and 1502 of the US Code of Federal Regulations that govern the implementation of the National Environmental Policy Act (NEPA).

Downtown to Downtown

The proposed maglev would have only three stops: downtown Washington, downtown Baltimore, and Baltimore/Washington International (BWI) Thurgood Marshall Airport. The present paper initially examines travel between the two urban centers and subsequently examines travel from both urban centers to BWI.

The maglev's draft environmental impact statement says that most of the maglev's ridership would be people traveling between the two cities rather than people flying out of or into BWI.

In addition, the draft environmental impact statement says that most maglev trips would be "diverted" rather than "induced." A diverted maglev trip is one that the customer would make by another form of transportation if the maglev were not built. In contrast, an induced maglev trip is one that would only occur if the maglev were built. As a practice, transportation planners divide total ridership into diverted and induced travel. The present chapter examines only diverted trips because they are easier to estimate than induced trips.

The calculation of diverted trips starts with a recent travel survey. The survey states how many trips are made between Washington and Baltimore. The survey was published in 2020 by the Metropolitan Washington Council of Governments.[8]

The relevant number to extract from the travel survey is the number of trips within the maglev service area: 18,956 one-way trips per day. As discussed in this chapter's appendix, this number depends on which jurisdictions are determined to be within the maglev's ridership area. These jurisdictions are identified in Chapter 3 (page 41). In these jurisdictions, most residents could save time by riding the maglev rather than driving between Baltimore and Washington. In this way, the maglev would serve three jurisdictions at the

---

6. The maglev DEIS public comment period was January 23 through May 24, 2021: Maryland Transit Administration (MTA) 17 March 2021, press release, https://www.mta.maryland.gov/articles/304. Redacted copy of the 2018 Louis Berger ridership report released on April 23, 2021, at https://bwmaglev.info/index.php/project-documents/deis#ridership-studies.

7. Eccleston 2014, pg. 257–258. The cited regulations can be found at https://www.ecfr.gov.

8. See the appendix of the present chapter for details about the Regional Travel Survey.

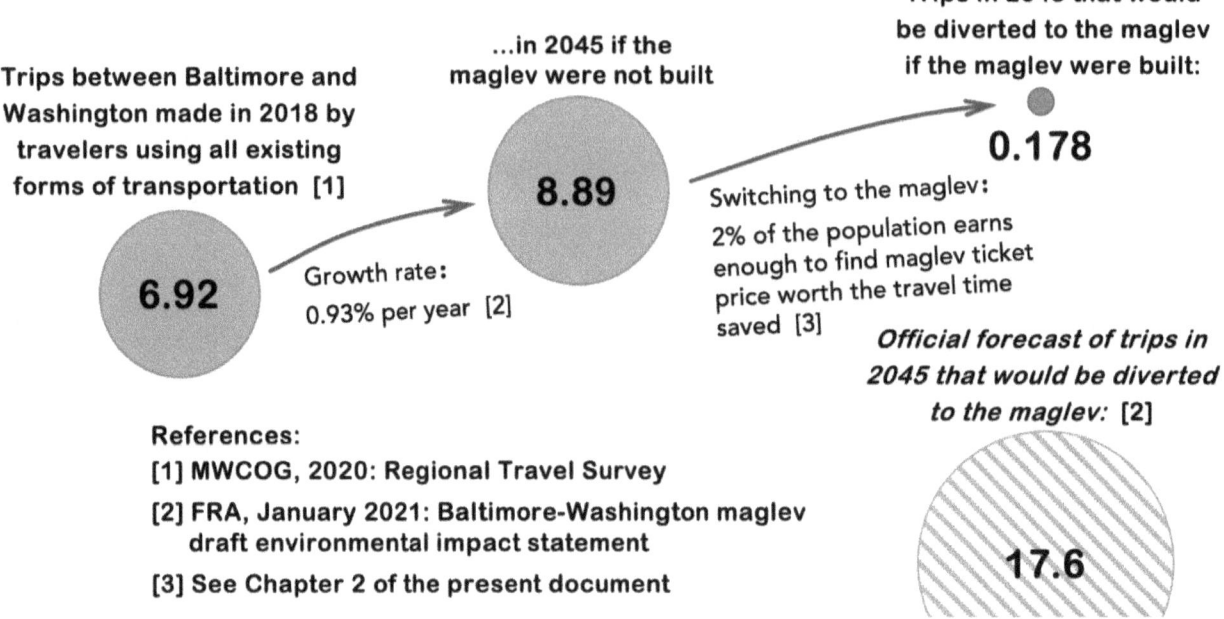

Figure 1. A schematic diagram showing how the present chapter calculates an unofficial forecast for the number of maglev trips in 2045 that would be made by "diverted" travelers. A diverted traveler is someone who would make the trip by another form of transportation if the maglev were not built. The 17.6-million-trip official forecast is much larger than the 178,000-trip unofficial forecast. Both forecasts exclude BWI airport passengers.

southern end of the maglev line: the District of Columbia, the City of Alexandria, and Arlington County. The maglev would serve two jurisdictions at the northern end of the line: the City of Baltimore and Baltimore County.[9]

The 18,956-trip estimate is based on data collected in 2018, but this number can be extrapolated to 2045, the year for which the maglev's official ridership forecast is intended to apply. To extrapolate from 2018 to 2045, one may use a 0.93% increase in travel per year between Baltimore and Washington as proposed in the maglev's draft environmental impact statement.[10]

The next step is to multiply by the fraction of the population that earns enough that the travel time saved on the maglev would seem worth the ticket price. In Chapter 2 of the present document, the author shows that about 2% of the population earns this much (page 27).[11]

---

9. See Chapter 3 of the present document.

10. 0.93% annual growth: Appendix D2, pg. C-106.

11. See Chapter 2 of the present document.

Figure 1 shows how these factors are combined to arrive at an unofficial forecast of 178,000 one-way trips diverted to the maglev in 2045. Figure 1 also shows the official forecast for this portion of the maglev ridership: 17.6 million one-way maglev trips. To be clear, both the official forecast and the just-derived unofficial forecast are both forecasts for diverted maglev trips in 2045, excluding BWI airport customers. The official forecast is approximately one hundred times greater than the independent, unofficial forecast (100 ≈ 17.6 ÷ 0.178).[12]

Downtown to Airport

The preceding section considered non-airport travel and this section considers airport travel. In both cases, the official ridership forecast in the draft environmental impact statement is much higher than the unofficial forecast derived in this chapter.

A maglev station is proposed immediately adjacent to BWI airport's main terminal where the hourly parking garage now stands. The Maryland Aviation Administration reported that BWI airport had 26.933 million arrivals and departures in 2019.[13]

The first task is to determine what portion of BWI customers would save time if they used the maglev to travel to or from the airport. Those Washington area residents who would save time riding the maglev to or from BWI are those who live in DC, Arlington, or Alexandria. Most City of Baltimore residents, but not most Baltimore County residents, could save time by riding the maglev to BWI. Approximately 21% of the region's population lives in the four above-mentioned jurisdictions.[14]

Next, apply to the airport trips the same two factors that were applied in the previous section to non-airport trips. The first factor extrapolates the 2019 measured trips to 2045, the year of the official maglev ridership forecast. The second factor is 0.02, the portion of the population wealthy enough to find the maglev travel-time savings worth the maglev ticket price.

After combining these factors, the result is an unofficial forecast of 143,000 one-way maglev trips in 2045 by BWI customers traveling to or from the airport on the proposed maglev. Add these 143,000 airport trips to the 178,000 non-airport trips derived in the previous section to arrive at the total number of maglev trips that represent travel diverted from other forms of transportation in 2045. The sum of these two numbers is 321,000 trips, which is far less than the official forecast of 20.6 million diverted trips.

To be clear, the official and unofficial forecasts are both estimates of the number of diverted maglev trips that would be made in 2045. The official forecast is a factor of 64 times greater than the independent, unofficial forecast that the present chapter derives (64 ≈ 20.6 ÷ 0.321).[15]

---

12. The official ridership forecast for diverted non-airport travelers is the product of 20.579 million trips by diverted travelers (Chapter 4.2, Table 4.2-3, pg. 4.2-7) multiplied by 85.5%, the percent of maglev trips that would be made by people other than BWI airport customers (Appendix D4, Table D4-19, pg. D-35). 17.6 million = 20.579 million · 0.855.
13. MD Aviation Administration December 2020.
14. 21%: see the Census Bureau data described in the appendix of the present chapter.
15. Official forecast of 20.579 million trips by diverted travelers: Chapter 4.2, Table 4.2-3, pg. 4.2-7.

Commuters

The official ridership forecast is far too high based on the analysis presented so far that uses publicly available reference datasets. Because it is a serious charge to claim that the Federal Railroad Administration has been fooled into republishing a grossly implausible ridership forecast, this section examines yet another reference dataset. This third dataset confirms the pattern seen so far, as explained below.

Data from the Census Bureau show that 13,091 people commuted between Baltimore and Washington in 2015, the most recent year for which data are available. This number is the sum of the people who live in Baltimore and work in Washington and those who live in Washington and work in Baltimore. As discussed on page 23 of the present document, these commuters have the District of Columbia, Arlington, or Alexandria at the southern end of their commute and Baltimore County or the City of Baltimore at the northern end.[16]

The annual number of one-way commuting trips can be estimated by multiplying the number of commuters by two trips per workday and by the average number of workdays in a year.[17]

Multiply this number of trips by the same two factors used in the previous sections of the present chapter. First, use a 0.93%-per-year increase in travel between the year that the data was collected, which was 2015, and the maglev forecast year, which is 2045. Second, multiply by 0.02 because only about 2% of the population is wealthy enough to make the travel time saved on the maglev worth the ticket price. The result is an unofficial forecast that 147,000 one-way maglev trips would be made in 2045 by diverted commuters, i.e., commuters who switch from other forms of transportation to ride the maglev.

In contrast, the official forecast is that diverted commuters would make 5.2 million one-way maglev trips per year. The official forecast is 35 times higher than the unofficial forecast ($35 \approx 5.2 \div 0.147$).[18]

To summarize, the present chapter has examined three reference datasets. All three provide evidence that the official ridership forecast for the proposed maglev is implausibly high. The official forecast in the draft environmental impact statement is more than ten times higher than the reference datasets can support.

Prior Studies Suggest Low Ridership

There is nothing surprising about the present chapter finding that only a few travelers would prefer the proposed Baltimore-Washington maglev over other forms of transportation.

A high-speed rail line shorter than 100 miles cannot compete with car travel according to a 1991 National Academies report and Federal Railroad Administration reports in 1993 and 2005. This result applies to all types of high-speed rail whether or not they use maglev technology. The proposed Baltimore-Washington maglev would be

---

16. 13,091 commuters: See the American Community Survey data in the appendix of the present chapter.

17. How many trips the average commuter would make in a year: appendix of the present chapter.

18. The official forecast for diverted commuters is the product of 20.579 million trips by diverted travelers (Chapter 4.2, Table 4.2-3, pg. 4.2-7) multiplied by 25.4%, the percentage of maglev trips that would be made by commuters (Appendix D4, Table D4-19, pg. D-35). 5.2 million = 20.579 million · 0.254.

only 36 miles long, which is much shorter than the 100-mile cutoff.[19]

It is surprising that the Federal Railroad Administration chose not to mention the findings of these earlier studies in the January 2021 draft environmental impact statement for the proposed Baltimore-Washington maglev. The regulations that implement the National Environmental Policy Act (NEPA) require that an impact statement evaluate all relevant points of view.[20]

The most natural interpretation of these earlier studies is that a maglev shorter than 100 miles would not be economically viable. For this reason, a short-run maglev line may be an invalid subject for an environmental impact statement. To quote NEPA regulations, the subject of an environmental impact statement must:

> have independent utility or independent significance, i.e., be usable and be a reasonable expenditure even if no additional transportation improvements in the area are made [21]

Based on the analysis in the present chapter, a maglev between Baltimore and Washington would have so few riders that it would lack independent utility.

Furthermore, high-speed rail service already exists between Baltimore and Washington. Specifically, this trip is one segment of Amtrak's existing Acela service from Washington to Boston. In May 2021, Amtrak announced that an Acela trip from Baltimore to Washington would take only 21 minutes after planned track improvements and a tunnel replacement project are completed. Therefore, the Acela trip length would be only somewhat longer than the advertised 15-minute trip length for the proposed Baltimore-Washington maglev.[22]

Raising the Alarm

Elected officials at the local, county, and state levels have already raised the alarm that the maglev would not be worth constructing because it would likely attract so few riders. For example, the City of Greenbelt stated in 2021:

> Of particular concern, the DEIS relies on undisclosed methodologies to predict wildly inflated ridership figures and savings in travel time. Based on reasonable ridership assumptions, it is unlikely the SCMAGLEV would be profitable. (page 4)

The City of Greenbelt also asserted on pages 22–25 of its report that the implausibly high ridership forecast is one reason why the maglev's draft environmental impact statement appears to violate National Environmental Policy Act (NEPA) regulations.

---

19. 33–36-mile length stated in the maglev DEIS: FRA 2021, Chapter 3, pg. 3-18 and 3-19. National Academies 1991, Figure ES-1, pg. 7. Car's advantages over rail: FRA 1997, pg. 7-4; FRA 2008, pg. 6-7; and FRA 2005, pg. ES-3.

20. Eccleston 2014, pg. 258–259. NEPA regulation 2005 Section 1502.9(a) states, "the [lead author] agency shall make every effort to disclose and discuss at appropriate points in the draft statement all major points of view on the environmental impacts."

21. 23 CFR § 771.111, https://www.law.cornell.edu/cfr/text/23/771.111.

22. Flynn (6 May 2021) testimony before the US House of Representatives: "[The] Maglev's projected trip time from Washington to Baltimore would be only 15 minutes faster than an Acela train today, and just six minutes faster than the projected Acela trip time following replacement of the B&P [Baltimore and Potomac] Tunnel and completion of the other investments discussed above."

In 2021, the Prince George's County Council stated that it opposed the construction of the maglev because of a "lack of usage access," i.e., too few county residents would ride it.[23]

In 2018, the District 22 delegation to the Maryland state legislature expressed its concerns about maglev ridership in the following way:

> We are writing to express our strong opposition to the proposed SCMAGLEV Project. To date, we are far from convinced that the SCMAGLEV Project is necessary, environmentally sound, financially sustainable or that a market exists outside of deep-pocketed corporate executives.

The co-signers were three members of the Maryland House of Delegates and a member of the Maryland Senate.[24]

Numerous environmental organizations support public transit and strongly oppose the proposed Baltimore-Washington maglev. For example, the Chesapeake Bay Foundation wrote in 2021:

> despite the overly generous ridership figures developed, we sincerely find it hard to fathom how the train can "meet the capacity and ridership needs" in the region and the generalized corridor if it will only make one stop between its two terminal stations, and if a one-way ticket average fare was projected in this study to cost at least $60 (in 2020).

In addition, 52 environmental organizations co-signed a letter in 2021 opposing the proposed Baltimore-Washington maglev for similar reasons.[25]

Both politically-aligned and non-partisan organizations and individuals have expressed concern that the proposed maglev would not attract sufficient ridership to make it worth building. The CATO Institute, a libertarian think tank, published the following comments by transportation analyst Randal O'Toole:

> Clearly, the main users of the maglev line will be bureaucrats and lobbyists who will have someone else (mainly taxpayers) pay their way. What is less clear is why ordinary taxpayers should pay to build a line that they won't ever use [26]

On the other side of the political spectrum, Martin Mitchell, the president of the Prince George's County Young Democrats, stated:

> It's obviously not going to be affordable to a lot of people so I don't understand how you expect to take a lot of cars off the road and be an alternative to driving [27]

Two non-partisan organizations, Citizens Against The SCMaglev and the Maryland Coalition for Responsible Transit, jointly published a document in 2021 that said the following about maglev ridership information in the draft environmental impact statement (DEIS):

---

23. PG County Council, Resolution CR-26-2021.

24. District 22 delegation (2018).

25. Chesapeake Bay Foundation (24 May 2021); 52 organizations: National Park Conservation Association et al. (24 May 2021).

26. Randal O'Toole, 6 April 2021: Maglev to destroy habitat, climate. *Cato At Liberty* blog, https://www.cato.org/blog/meglev-destroy-habitat-climate. O'Toole is also the author of the book *Romance of the Rails: Why the Passenger Trains We Love Are Not the Transportation We Need* (CATO Institute, 2018).

27. Truss-Williams 22 April 2021.

The DEIS fails to provide the financial, ridership, job creation, and other required data and analyses needed to substantiate the Project sponsor's claims about the benefits and viability of their financial model and forecasts (page 9)

Conclusion

The present chapter has examined the official forecast for the number of trips that would be made on the proposed Baltimore-Washington superconducting maglev. The official forecast is stated in the draft environmental impact statement published by the Federal Railroad Administration in January 2021.

The analysis in the present chapter finds that the official ridership forecast is implausibly high. The official forecast is more than an order of magnitude higher than what reference datasets can support.

The official forecast is that 20.6 million one-way maglev trips would be made each year by travelers diverted from other forms of transportation. In contrast, various reference datasets examined in the present chapter suggest that a much smaller number of diverted travelers is more likely: 0.32 million one-way maglev trips per year. A diverted traveler is someone who would make the trip by another form of transportation if the maglev were not built. The draft environmental impact statement reports that the great majority of maglev travelers would be diverted from other forms of transportation.

If the official ridership forecast is higher than warranted, then it would prevent the draft environmental impact statement from helping the public and elected officials evaluate the harm and benefits of the proposed maglev. Ridership influences, among other things, the maglev's revenue, the solvency of the maglev operator, air-pollution reduction, road-congestion improvement, and jobs created by maglev operations.

References

Bureau of Labor Statistics, 2020: *National Compensation Survey: Employee Benefits in the United States, March 2020*. Bulletin 2793, https://www.bls.gov/ncs/ebs/benefits/2020/employee-benefits-in-the-united-states-march-2020.pdf.

Chesapeake Bay Foundation, 24 May 2021: RE: Draft Environmental Impact Statement and Draft Section 4(f) Evaluation Baltimore-Washington Superconducting MAGLEV Project. comment on draft environmental impact statement, 19 pp, obtained through the office of Delegate Geraldine Valentino-Smith, Maryland House of Delegates.

CATS and MCRT, 20 May 2021: *SCMAGLEV DEIS Comments, Concerns, and Questions*. Citizen's Against the SCMAGLEV (CATS) and Maryland Coalition for Responsible Transit (MCRT), 495 pp, http://www.mcrt-action.org/. Click on "SCMAGLEV Opposition" link.

District 22 delegation, 18 January 2018: Letter to SCMAGLEV Project. 1 pg., https://www.greenbeltmd.gov/maglev/. The individuals signing the letter were Maryland State Senator Paul Pinsky and Delegates Tawanna Gaines, Anne Healey, and Alonzo T. Washington.

Eccleston, C. H., 2014: *The EIS Book*. CRC Press, 472 pp.

Federal Railroad Administration, January 2021: *Baltimore-Washington Superconducting MAGLEV Project Draft Environmental Impact Statement and Draft Section 4(f) Evaluation*. 3,053 pp. (main text and appendices), https://bwmaglev.info/index.php/project-documents/deis.

Federal Railroad Administration, 2008: *Analysis of the Benefits of High-Speed Rail on the Northeast Corridor*. Report CC-2008-091, memorandum from D. Tornquist, 19 pp, https://www.oig.dot.gov/library-item/30401.

Federal Railroad Administration, 2005: *Report to Congress: Costs and Benefits of Magnetic Levitation*. 76 pp, https://railroads.dot.gov/elibrary/report-congress-costs-and-benefits-magnetic-levitation.

Federal Railroad Administration, 1997: *High-Speed Ground Transportation for America*. 182 pp, https://railroads.dot.gov/sites/fra.dot.gov/files/fra_net/1177/cfs0997all2.pdf.

Federal Railroad Administration, September 1993: *Final Report on the National Maglev Initiative (NMI)*. Technical Report DOT/FRA/NMI-93/03. 121 pp, https://railroads.dot.gov/elibrary/final-report-national-maglev-initiative.

Flynn, W. J., 6 May 2021: Testimony before the US House of Representatives Committee on Transportation Infrastructure Subcommittee on Railroads, Pipelines, and Hazardous Materials. Hearing title: *The Benefits and Challenges of High-Speed Rail and Emerging Rail Technologies*, 20 pg., https://transportation.house.gov/imo/media/doc/Flynn%20Testimony2.pdf. Quoted by Lazo (6 May 2021, *Washington Post*).

Greenbelt, City of, 24 May 2021: *Comments by the City of Greenbelt*. Comments on the *Baltimore-Washington Superconducting Maglev Project Draft Environmental Impact Statement and Department of Transportation Act Section 4(f) Evaluation*. These comments were also adopted by the City of College Park and the Town of Landover Hills, 222 pp, https://www.greenbeltmd.gov/maglev/.

Hankey, S., G. Lindsey, and J. Marshall, 2014: Day-of-year scaling factors and design considerations for nonmotorized traffic monitoring programs. *Transportation Research Record: Journal of the Transportation Research Board*, No. 2468, National Academies, 64–73.

Hartgen, D. T., 2013: Hubris or humility? Accuracy issues for the next 50 years of travel demand modeling. *Transportation*, doi: 10.1007/s11116-013-9497-y. http://hartgengroup.net/Projects/National/USA/hubris_humility/2013-08-28_FINAL_PAPER_OnLine%20Transportation_40.6_Sept_2013.pdf.

Louis Berger, 08 November 2018: *Baltimore-Washington SCMAGLEV Project Final Ridership Report*. Section 2.2 "Document Travel Demand," 79 pp, https://bwmaglev.info/index.php/component/jdownloads/?task=download.send&id=71&catid=6&m=0&Itemid=101. This copy of the report is heavily redacted and was publicly released on April 23, 2021.

Maryland Aviation Administration, December 2020: *Monthly Statistical Report Summary for the Month of December 2020*. 17 pp, https://www.bwiairport.com/sites/default/files/Dec2020.pdf, cited in https://www.bwiairport.com/flying-with-us/about-bwi/statistics. Calendar year 2019 number of commercial passengers: 26,933,896 (page 5).

MWCOG, 2019: *2019 State of the Commute Survey: Technical Survey Report*. Commuter Connection Program of the National Capital Region Transportation Planning Board (NCRTPB) part of the Metropolitan Washington Council of Governments, 219 pp. States on pages ii and 8 that teleworking reduces by 9.7% the number of commuters trips in a typical weekday in the Washington region.

MWCOG, 2020: Regional Transportation Data Clearinghouse (RTDC) 2017/2018 Regional Travel Survey (RTS) Tabulations. comma-separated-value *.csv files, National Capital Region Transportation Planning Board (NCRTPB) part of the Metropolitan Washington Council of Governments. Data files downloadable from https://rtdc-mwcog.opendata.arcgis.com/datasets/regional-travel-survey-rts-tabulations. Introductory page: https://www.mwcog.org/transportation/data-and-tools/household-travel-survey/.

Jon, K., 21 January 2021: *2017-2018 Regional Travel Survey Briefing: Change in Observed Trips Since 2007/08*. technical presentation, Transportation Planning Board (NCRTPB) part of the Metropolitan Washington Council of Governments, https://www.mwcog.org/documents/2020/01/21/regional-travel-survey-presentations-regional-travel-survey-tpb-travel-surveys/.

National Academies of Science, Engineering, and Medicine, 1991: *In Pursuit of Speed: New Options for Intercity Passenger Transport--Special Report 233*. The National Academies Press, https://doi.org.10.17226/11408, 185 pp.

National Park Conservation Association and 51 environmental organizations, 24 May 2021: comments on maglev DEIS, 5 pp. Obtained from the Maryland Coalition for Responsible Transit (MCRT).

Prince George's County Council, 18 May 2021: *A resolution concerning the Baltimore-

Washington Superconducting Maglev Project - Opposition. Resolution CR-26-2021, https://princegeorgescountymd.legistar.com/LegislationDetail.aspx?ID=4837039&GUID=B63C908A-0CFF-4C5B-8453-EF6EB5C531DF.

Truss-Williams, A., 22 April 2021: Community members say MAGLEV train would be overpriced, destroy local environments. *The Diamondback*, the independent student newspaper of the University of Maryland, College Park, MD, https://dbknews.com/2021/04/22/community-members-say-maglev-train-would-be-overpriced-destroy-local-environments/.

Voulgaris, C. T., 2019: Crystal balls and black boxes: what makes a good forecast? *J. Planning Literature*, **34**, 286–299, https://doi.org/10.1177/0885412219838495.

US Census Bureau, 2015: Table 4, Residence MCD/County to Workplace MCD/County Commuting Flows for the United States and Puerto Rico Sorted by Workplace Geography: 5-Year ACS, 2011–2015. An Excel spreadsheet for the entire country with over 594,000 rows. On the web page titled "2011–2015 5-year ACS commuting flows," https://www.census.gov/data/tables/2015/demo/metro-micro/commuting-flows-2015.html.

US Census Bureau, 2006: *Current Population Survey Design and Methodology*. technical paper 66, 175 pp, https://www.census.gov/prod/2006pubs/tp-66.pdf.

Appendix

This appendix describes the official and unofficial forecasts for various categories of ridership for the proposed Baltimore-Washington maglev. The unit of ridership is a one-way trip on the maglev made by one person, regardless of which maglev stations the passenger uses. The official ridership forecast is extracted from the maglev's draft environmental impact statement (DEIS), as summarized in Table 1 of the present chapter.

Table 1. The official ridership forecast for the proposed Baltimore-Washington maglev as published in the January 2021 draft environmental impact statement.

Category of maglev traveler	One-way maglev diverted trips in 2045	
	Percent of diverted trips	Number of diverted trips
1. Maglev, diverted non-airport	85.5%[b]	17.595 million
2. Maglev, diverted airport customers	14.5%[b]	2.984 million
3. Maglev, diverted (non-airport + airport)[a]	100%	20.579 million
4. Maglev, diverted commuter (a subcategory of diverted non-airport customers)	25.4%[b]	5.227 million
5. Maglev, total ridership (diverted + induced)[a]	---	24.939 million

[a] Diverted and total ridership are stated explicitly in the draft environmental impact statement: Chapter 4.2, Table 4.2-3, pg. 4.2-7. The quoted numbers are for the year 2045 if the downtown Baltimore maglev station was located at Camden Yards.

[b] Percent commuters and percent airport: Appendix D4, Table D4-19, pg. D-35. The number of one-way trips for these rows are calculated using the number of trips in row 3 (20.579 million trips) multiplied by these percentages.

The unofficial forecast that is derived in the present chapter is based on the various reference datasets described in this appendix. The unofficial forecast is stated in Table 2 of the present chapter. Tables 3 through 5 summarize the reference datasets that support the unofficial forecast.

It is unclear whether any of the reference datasets used to create the unofficial ridership forecast are also used in the DEIS to create the official ridership forecast. The DEIS is vague about its input data. On this topic, the most precise sentence in the DEIS is the following vague sentence:

> A comprehensive accounting of current intercity trips was developed utilizing MPO surveys and models, transit agency data, airport data, and mobile phone origin/destination data.

The Official and Unofficial Forecasts

The present chapter discusses one portion of the maglev's ridership exclusively. Specifically, it focuses on maglev travelers that are diverted from other forms of transportation. A diverted traveler is someone who would make the trip using the maglev if it were built and who would make the trip by another form of transportation if the maglev were not built. The DEIS contains a forecast for the maglev's diverted ridership in 2045 under the assumption that the downtown Baltimore maglev station would be built at Camden Yards.

The DEIS's official ridership forecast is shown in the rightmost column of Table 1 of the present chapter. The total number of diverted trips is 20.579 million one-way maglev trips in 2045 as stated in the DEIS, Chapter 4.2, Table 4.2-3 (page 4.2-7). In contrast, the unofficial forecast for the same quantity is only 0.321 million one-way maglev trips in 2045, as stated in Table 2 of the present chapter. The official forecast is approximately 64 times greater than the unofficial forecast. These forecasts are so different that both cannot be collect.

One way that the DEIS categorizes diverted travelers is that they are either airline customers at BWI airport or travelers with another, non-airport, travel goal. The DEIS identifies commuters as a subcategory of non-airport diverted travelers.

The DEIS does not explicitly state the ridership (trips per year) for these specific categories of diverted travelers. But the ridership can be easily calculated from the percentages stated in the DEIS's Appendix D4, Table D4-19 (page D-35). These percentages are used to populate rows 1, 2, and 4 of Table 1 of the present chapter.

An Unofficial Forecast of Non-Airport Diverted Travelers

The unofficial forecast for non-airport diverted travelers is 178,000 one-way maglev trips in 2045, as stated in the rightmost column of Table 2. To derive this number, start with data from the Regional Travel Survey that was published in 2020 based on data collected in 2017 and 2018. This survey was developed by the National Capital Region Transportation Planning Board (NCRTPB), an organization within the Metropolitan Washington Council of Governments, MWCOG. (See also Jon 2021.)

On page 23, Table 3 gives the relevant data from the Regional Travel Survey. The total number of one-way trips between the cities is 18,956 during a typical weekday. As a simplifying assumption, the present chapter assumes that average traffic on a Saturday or Sunday is equal to average traffic on a weekday, Monday through Friday. For this reason, the daily trip count can be multiplied by the number of days in a year (365.25 days) to obtain an estimate of the annual trip count in 2018 (6.924 million trips per year).

Table 2. The unofficial ridership forecast for the proposed Baltimore-Washington maglev as derived in the present chapter.

Category of traveler	All modes of transportation, assuming the maglev is not built		Trips diverted to the maglev
	One-way trips before diversion [baseline year]	One-way trips in 2045, before diversion [d]	One-way diverted trips in 2045 [e]
1. Non-airport	6.924 million [2018] [a]	8.890 million	0.178 million
2. Airport customers	5.637 million [2019] [b]	7.170 million	0.143 million
3. Total diverted trips including non-airport and airport	---	---	0.321 million
4. Commuter (a subcategory of non-airport customers)	5.577 million [2015] [c]	7.362 million	0.147 million

[a] MWCOG Regional Travel Survey (RTS) published in 2020 using data collected in 2018. For details see Table 3 of the present chapter.

[b] The Maryland Aviation Administration reports 26.933 million passengers passed through Baltimore/Washington International (BWI) Thurgood Marshall airport in 2019. According to the US Census data shown in Table 4 of the present chapter, 20.93% of the region's residents live in the District of Columbia, Arlington, Alexandria, or City of Baltimore. 5.640 = 0.2093 x 26.933.

[c] American Community Survey (ACS) data from the Census Bureau as display in Table 5 of the present chapter.

[d] The scale factor fgrowth to account for an 0.93%-per-year growth in travel in the region. The factor f_{growth} equals 1.284, 1.272, or 1.320 for a start year of 2018, 2019, or 2015 and an end year of 2045. See the equation that defines f_{growth} in the appendix of the present chapter.

[e] This column is calculated by multiplying the 2045 trips before diversion by 0.02 based on Chapter 2 on page 27 of the present document.

This is the number of trips in 2018 without the maglev, obviously, because the proposed maglev didn't exist in 2018.

This simplifying assumption about weekend-vs.-weekday traffic volume is acceptable for approximate calculations. For tourist, recreation, and leisure-activity destinations, weekend traffic may be greater than weekday traffic. Conversely, an office dense location may have much less weekend traffic than weekday traffic. Hankey et al. (2014, Table 2) found that average traffic volume on weekend days vs. weekdays varies typically by approximately ±20% depending on the primary use of that location.

The decision of which jurisdictions to extract from the Regional Travel Survey is based on Chapter 3 of the present document (page 41). At the northern end, Baltimore County and the City of Baltimore are included in this ridership area. To work with entire jurisdictions, Anne Arundel

County is not included in the maglev ridership area. Most Anne Arundel County residents would find it faster to drive directly to destinations in the Washington area rather than detour northward to reach the maglev station at BWI or downtown Baltimore.

Presumably, BWI airport passengers would be a small portion of the trips included in the Regional Travel Survey's total for the number of trips with one end in the District of Columbia, Alexandria, or Arlington and the other end in Anne Arundel County. It is impossible to know based on the Regional Travel Survey the exact portion of trips made by BWI airport passengers. For this reason, the Regional Travel Survey is not well suited for analyzing airport customers. Airport customers are treated in the next section of the appendix using a different reference dataset.

The simplest formula to extrapolate 2018 non-airport travel data to 2045 is to use a fixed annual increase in traffic:

$$f_{growth} = \{ 1 + (a \div 100\%) \}^d$$

where d is the number of years, a is the percent increase in traffic in a single year, and f_{growth} is the fractional increase in traffic over the stated period of years. Initially, consider travel growth during a 27-year period (2018 to 2045) at the annual growth rate of 0.93% that was specified in the maglev DEIS. Under these conditions, traffic volume would grow by a factor of 1.284 between 2018 and 2045. Applying this growth factor, one obtains 6.924 million one-way trips in 2045 made by all forms of transportation, assuming the maglev were not built. The result is the bottom number in Table 3 of the present chapter.

The last step is to convert the number of trips between Baltimore and Washington in 2045 by non-maglev forms of transportation into the forecasted portion of these trips that would be diverted to the maglev were it built. Two percent of these trips would be diverted to the maglev according to Chapter 2 (page 27). The result is an unofficial forecast for the number of non-airport diverted trips on the maglev in 2045: 178,000 one-way maglev trips, as stated in Table 2 of this chapter.

An Unofficial Forecast of Airport Diverted Travelers

The number of passengers arriving and departing BWI airport is reported each year. The organization that runs the airport is called the Maryland Aviation Administration. According to the administration, the number of passengers was 26.933 million in 2019. Each one of these airplane trips necessitates one passenger trip to or from the airport.

The question is what fraction of the total number of BWI trips (26.933 million) in the entire region would occur specifically within the maglev ridership area. One might assume that the fraction of trips would be approximately the same as the fraction of the region's population that lives within the maglev ridership area.

This fraction of the region's population can be calculated from Census Bureau population data provided in Table 4 on page 25. The percent of the region's population that lives in the jurisdictions where the maglev could save them time on a trip to BWI airport is 20.93%, based on this data. Traveling south on the maglev, only City of Baltimore residents would travel to BWI. In contrast, most Baltimore County residents would find it just as fast to drive directly to BWI and skip the unnecessary expense of the maglev, according to Chapter 3 of the present document. At the southern end of the maglev line, the residents of the District of Columbia, Arlington, and Alexandria would save time riding the maglev north to BWI.

Multiplying the 2019 number of BWI trips by 0.2093 gives 5.637 million BWI trips in 2019 in the maglev's ridership area. Multiplying this number by 1.272 projects this number 26 years into the future to 2045, using the formula for f_{growth} that was introduced in the previous section of the appendix. Multiplying this number by 0.02 converts the total number of trips assuming the maglev isn't built into the number of trips that would be diverted if the maglev were built. The 0.02 factor comes from the previous section of the appendix. The result is an unofficial forecast of 143,000 one-way maglev trips in 2045 made by BWI-airport customers traveling to or from the airport. This number is stated in Table 2.

An Unofficial Forecast of Diverted Commuters

The American Community Survey (ACS) of the Census Bureau provides a travel matrix of commuters. In other words, the ACS tabulates the number of people living in a particular US county or city and commuting to work in various other counties or cities. The most recent year for which these data are available is 2015. Table 5 of this chapter shows the relevant ACS data for estimating diverted commuter traffic on the proposed Baltimore-Washington maglev.

Table 3. Trips between Baltimore and Washington on a typical weekday as estimated by the Metropolitan Washington Council of Governments' Regional Travel Survey published in 2020 [a]

Direction of travel	Northern terminus of trip		Total [b]
	City of Baltimore	Baltimore County	
Northbound [c]	6,033 ± 973	3,232 ± 681	9,265 ± 1,188
Southbound [d]	6,898 ± 1,067	2,793 ± 675	9,691 ± 1,263
Total number of trips during typical weekday in 2018:			18,956 ± 1,733
Total number of trips per year in 2018 [e]:			6.924 ± 0.633 million

[a] The southern terminus of each trip is in the District of Columbia; Arlington County, Virginia; or the City of Alexandria, Virginia.

[b] The standard deviation of a sum is calculated here as the square root of the squares of the standard deviations of the individual terms of the sum: $\sigma_{x+y}^2 = (\sigma_x^2 + \sigma_y^2)^{1/2}$.

[c] Northbound trips have their northern terminus as the trip destination. The relevant RTS trip data for the core region (DC, Arlington, Alexandria) with trip terminus in Baltimore City or Baltimore County are found in the file called T09_D_STATE_COUNTY_FIPS_a and in the rows begin with "Activity Center."

[d] Southbound trips have their northern terminus as the trip origin. The relevant RTS trip data for the core region (DC, Arlington, Alexandria) with trip origin in Baltimore City or Baltimore County are found in the file called T05_O_STATE_COUNTY_FIPS_a and in rows that begin with "Activity Center."

[e] Assuming that the number of trips on a Saturday or Sunday is equal to those on a typical weekday (Monday through Friday), then an estimate of the annual number of trips can be generated by multiplying the value for a typical day by 365.25 d y^{-1}.

Using data in Table 5 of the present chapter, one can calculate the number of commuters in 2015 who traveled between Baltimore and Washington by summing over all existing forms of transportation. The southern end of the commute is in the District of Columbia, Arlington County, or the City of Alexandria. The northern end of the commute is in Baltimore County or the City of Baltimore. These jurisdictions were chosen based on Chapter 3 of the present document.

To convert the number of commuters to the number of annual trips made by commuters, one first needs to estimate the number of days in a year that the average person commutes. There are 261 workweek days in a year, i.e., Monday through Friday each week of the year (261 days ≈ 365.25 days · 5 ÷ 7). From these 261 days, subtract 10 federal holidays and subtract 3 weeks of paid vacation (15 workdays). The result is 236 workdays per year.

Because a fraction of the Washington region's workforce will telework from home on a typical workday, the number of travel-to-the-office days is less than the number of workdays for the average person. The Metropolitan Washington Council of Governments reported in 2019 that teleworking reduced the number of daily commutes to the office by 9.7% relative to the number that would have occurred without teleworking.

Multiply the 236 workdays by 0.903 because 9.7% of workdays are telework days in the Washington region (0.903 = {100% -9.7%} ÷ 100%). The result is that there are 213 days per year in which the average worker in the Washington region travels to the office (213 days = 236 days · 0.903). Assuming two trips during each of these days means that, in a year, each commuter in the Washington region makes on average 426 one-way trips to the office.

Multiply the number of commuters by 426 one-way trips per year, and the result is 5.577 million one-way trips between Baltimore and Washington in 2015 by all forms of available transportation, obviously excluding the maglev because it didn't exist in 2015. Next, multiply by the previously defined factor f_{growth} to extrapolate from 2015 to 2045. Multiply by 0.02 to convert from all modes of non-maglev transportation to the number of such trips that would be diverted to the maglev if it were built. The result is an unofficial forecast of 147,000 one-way maglev trips made by diverted commuters in 2045. This number is stated in Table 2.

Table 4. The population of counties and cities in the planning areas of the Metropolitan Washington Council of Governments (MWCOG) and the Baltimore Metropolitan Council (BMC).

Location	Population[a]
Washington-area jurisdictions served by the proposed maglev[b]	1,102,019
District of Columbia	705,749
Arlington County, VA	236,842
City of Alexandria, VA	159,428
Baltimore-area jurisdictions served by the proposed maglev[b]	1,420,860
City of Baltimore, MD	593,490
Baltimore County, MD	827,370
Jurisdictions not served by the proposed maglev	5,579,749
Fairfax County, VA	1,147,532
Prince William County, VA	470,335
Loudon County, VA	413,538
Frederick County, MD	259,547
Montgomery County, MD	1,050,688
Prince George's County, MD	909,327
Carroll County, MD	168,447
Howard County, MD	325,690
Ann Arundel County, MD	579,234
Hartford County, MD	255,411
Population served by the proposed maglev for travel between Baltimore and Washington	2,522,879
Population served by the proposed maglev for travel to Baltimore/Washington International (BWI) airport[b]	1,695,509[c]
Total population in the Baltimore-Washington region	8,102,628

[a] As of 2019 according to the US Census. Data in *.csv format: https://www2.census.gov/programs-surveys/popest/datasets/2010-2019/counties/totals/co-est2019-alldata.csv. Description: https://www.census.gov/data/datasets/time-series/demo/popest/2010s-counties-total.html.

[b] A jurisdiction is served by the maglev if the majority of its residents would save time riding the maglev rather than making the trip by car. See Chapter 3 on page 41 of the present document for details. The present chapter includes Baltimore County in the area served by the maglev for travel between Baltimore and Washington largely because many Baltimore County residents would save time if they used the maglev station at BWI. However, when the destination is BWI, then only the downtown Baltimore maglev station can serve as a starting point, which is too far out of their way for most Baltimore County residents to find useful.

[c] This population is 20.93% of the Baltimore-Washington region's population.

Table 5. Number of people commuting between Baltimore and Washington as estimated by the Census Bureau's American Community Survey (ACS) in 2015 [a]

Location of home	Job location					
	Baltimore		Washington			
	Baltimore City	Baltimore County	District of Columbia	Arlington	Alexandria	Total
Baltimore City			4,765	392	182	5,339
Baltimore County			5,120	369	247	5,736
District of Columbia	1,234	403				
Arlington	115	16				
Alexandria	231	17				
Total	1,580	436			Number of commuters:	13,091
					Number of one-way trips per year [b]:	5.577 million

[a] Data from the US Census Bureau, 2015: Table 4, Residence MCD/County to Workplace MCD/County Commuting Flows for the United States and Puerto Rico Sorted by Workplace Geography: 5-Year ACS, 2011–2015. An Excel spreadsheet for the entire country with over 594,000 rows. On the web page titled "2011–2015 5-year ACS commuting flows," https://www.census.gov/data/tables/2015/demo/metro-micro/commuting-flows-2015.html.

[b] The total number of trips is the number of commuters times 426 one-way trips per year. The number of one-way trips per year is 2 one-way trips per day of traveling to the office: 426 = 2 · 0.903 · 236. The factor of 0.903 comes from the fact that Washington-region workers spend, on average, 9.7% of their days teleworking rather than traveling to their office. The 236 figure is the number of weekday in the year minus 10 federal holidays and minus 3 weeks of paid vacation (15 days): 236 = 365.25 (5/7) - (10 + 15) = 261 -25.

2 • WEALTH

Most maglev riders would belong to the wealthiest 2% of the Baltimore-Washington population

An individual would have to earn at least $363,000 a year for him or her to find the maglev ticket price worth the modest reduction in door-to-door travel time

Common sense tells us that few people would be willing to pay 40 to 80 dollars to save just 8 to 27 minutes. For this reason, the advertised utility of a 17-billion-dollar project would shrink to almost nothing. The project in question is the proposed Baltimore-Washington superconducting magnetic-levitation rail line known as "the maglev."[1]

Here are the facts. In January 2021, the Federal Railroad Administration published the draft environmental impact statement for the proposed maglev. In this document, the agency stated that maglev customers would save on average 8 to 27 minutes of travel time, door to door. The agency also considered various options for the maglev's ticket price but settled on $40 to $80 in the computer simulation that was used to generate the official forecast of how many trips would be made on the maglev.[2]

Realistically, only a small proportion of the population is wealthy enough to pay this much to save so little time. US Census data and the calculation described below suggest that no more than 4% of workers in the Baltimore-Washington region earn this much. Two percent of workers is more likely.

Background

The companies that want to build a maglev between Baltimore and Washington are trying

---

1. $15–17 billion construction cost: DEIS, Appendix D4, Table D4-8, pg. D-21; 8–27 minutes saved travel time: Appendix D4, pg. C-6; $40–$80 ticket price: Appendix D2, pg. D-107, D-108.
2. The DEIS considered a maglev ticket price as low as $27 but determined the official ridership forecast based on a $40–$80 ticket price: Appendix D2, pg. D-107, D-108, "Final SCMAGLEV Fare Assumptions" section.

to persuade elected officials and the public that the proposed maglev is for everyone, not just the rich. These companies are Baltimore Washington Rapid Rail (BWRR) and its parent company, The Northeast Maglev. On its website, BWRR states that the maglev would be "highly valued across all travel purposes and income segments." The Northeast Maglev's website states that the company is "looking in to innovative ways to make the train accessible to all." [3]

The question of whether only the wealthy would ride the maglev is ignored in the Executive Summary of the draft environmental impact statement that the Federal Railroad Administration published in January 2021. Buried in an appendix is a limp sentence on this subject: "Higher income workers would be the most likely to use SCMAGLEV for commuting" (Appendix D4, pg. D-81). Then again, the draft environmental impact statement also implies that a majority of the region's residents would find the maglev a good value:

> The ridership report assumes that about 70.0 percent of business travelers in the defined catchment area and 67.0 percent of non-business travelers, which includes those making personal trips as well as commuters, between Baltimore and Washington, D.C. would choose the SCMAGLEV service if it were available. (Chapter 4.6, pg. 4.6-3)

> The anticipated SCMAGLEV services are estimated to reduce travel times by 8 to 27 minutes of travel time savings depending on the trip purpose and length under each of the Build Alternatives. (Appendix D4, pg. C-6)

These two quotes seem to imply that 67% of the population is wealthy enough that the maglev ticket price would seem worth saving just 8 to 27 minutes of travel time. If this is the meaning that the Federal Railroad Administration intended, then the reader's first reaction may be that 67% seems too high because a single maglev ticket would cost $40 to $80 one way.

Further complicating matters, it is unclear what the 67% refers to because its description in the draft environmental impact statement is so brief. The statement obtained the 67% figure from a ridership report written by the Louis Berger consulting company. The public cannot read this report because it is one of many documents that underlie the draft environmental impact statement that are hidden from public view. For all we know, even the Louis Berger report does not adequately explain the meaning and derivation of the 67% figure.

Giving the environmental review of the proposed maglev the air of a farce, the Federal Railroad Administration released a heavily redacted copy of the Louis Berger ridership report toward the end of the public-comment period for the draft environmental impact statement. This redacted copy is a mere shell, completely blanking out both numeric data and text that would assist in interpreting the 67% figure and other aspects of the maglev's ridership forecast. [4]

To generate a precise forecast for the fraction of the population that would use a proposed

---

3. BWRR: https://bwrapidrail.com; TNEM: northeastmaglev.com/.

4. Page 48 of Louis Berger (2018 Nov 08) states the 67% figure, according to the DEIS, Chapter 4.6, page 4.6-3, footnote 9. The FRA released a heavily redacted copy of the Louis Berger report at bwmaglev.info/index.php/project-documents/deis, on 23 April 2021. The DEIS comment period was January 23 through May 24, 2021: https://www.mta.maryland.gov/articles/304.

transportation facility, complicated analysis of carefully constructed surveys is required. It may involve "mode-choice analysis of stated-preference surveys," to quote the maglev's draft environmental impact statement.[5]

The calculation is much simpler if the goal is just an approximate upper and lower bound on the fraction of the population that would find the travel cost and time savings attractive. An approximate calculation is simple enough to perform on a handheld calculator instead of requiring simulation software designed by a team of experts.[6]

The calculation described in the present chapter provides this sort of reality check. Mathematical details and supporting data are provided in the chapter's appendix.

Serving the 2%

Before estimating who would ride the maglev, one needs to take care of two preliminaries. First, one chooses an estimate for how much a traveler would be willing to pay to save time. A plausible approximation is that an individual is willing to pay for travel-time savings at a rate similar to the rate at which he or she earns money at his or her job. Transportation models implement this basic idea in various ways.

Second, one needs an estimate for the averages of two quantities. These quantities are the price difference and the door-to-door travel-time difference between riding the maglev and driving directly to the destination. The maglev would be more expensive than driving and in some cases faster depending on the location of the trip origin and destination. A range for the travel-time difference is stated in the draft environmental impact statement: 8 to 27 minutes. Determining the price difference requires a little math.[7]

The price difference would be $33 to $73 for an individual traveling alone and much more for a family traveling together as discussed later. This estimate for an individual traveling alone comes from taking the $40-to-$80 per-person one-way maglev ticket price stated in the draft environmental impact statement and subtracting the cost of driving. The per-vehicle cost of driving a car between Baltimore and Washington is about $7, and this estimate can be calculated from two numbers. Start with the draft environmental impact statement, which states that typical car trips between the two cities are 39.6-miles long. Multiply that distance by AAA's estimate of a typical car's per-mile cost for gas and maintenance. One could use a somewhat different value than $7 for the cost of driving and the results would be essentially unchanged, as discussed in this chapter's appendix.[8]

The middle of the above-mentioned range for the extra cost to ride the maglev is $53, and the middle of the time-savings range is 17.5 minutes.

Someone who finds it a fair deal to pay about $53 to save about 17.5 minutes would demonstrate a willingness to pay $181.71 per hour. Such a person would most likely earn at least that much per hour, which would mean an annual income of

---

5. Appendix D2, pg. C-105.

6. This topic is discussed in Chapter 12 of Ortuzar and Willumsen (2011).

7. 8–27 minutes saved travel time: Appendix D4, pg. C-6.

8. 7.08 = 39.6 · 0.1787; 39.6 mile trip length: Appendix D4, Table D4-59, pg. E-82; $0.1787/mile for medium sedan: AAA 2020.

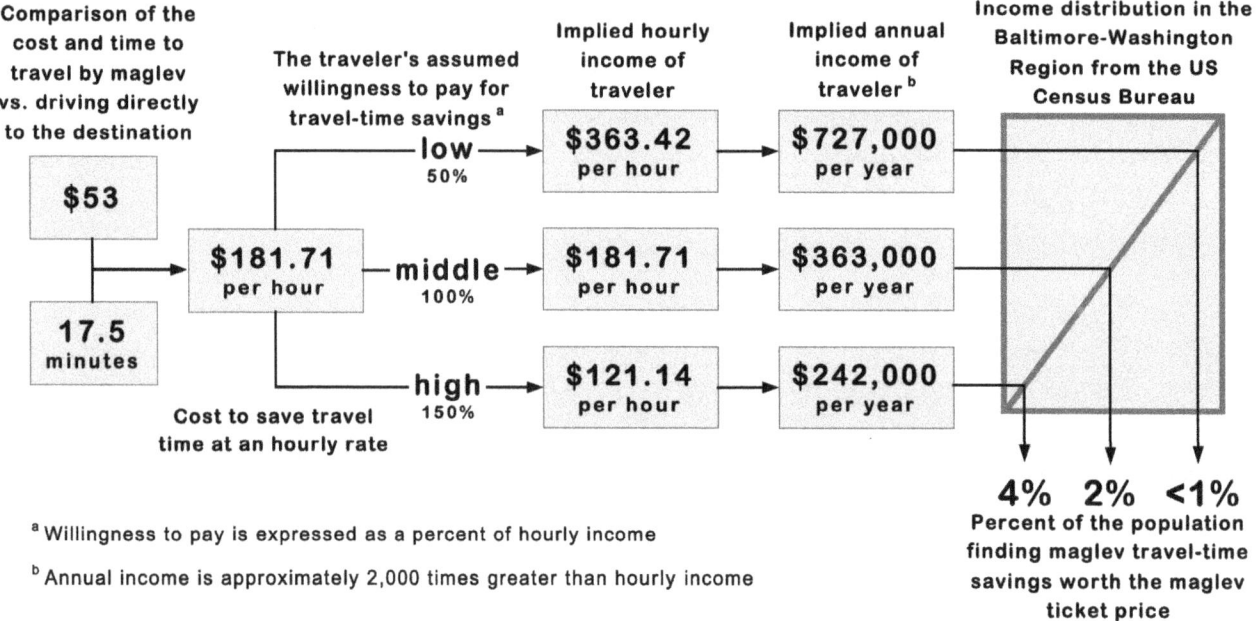

Figure 1. A schematic diagram showing a way to estimate the percentage of the population that would choose to ride the maglev given the average time savings and maglev ticket price. The percent calculated (approximately 2%) applies to an individual traveling alone, not a family traveling together.

about $363,000. Annual income is about 2,000 times greater than hourly income.[9]

The US Census Bureau reports that only about 2% of workers in the Baltimore-Washington region earn at least $363,000 a year. Therefore, we can conclude that only about 2% of workers would choose to ride the maglev. As discussed in this chapter's appendix, only 2% to 3% of workers earn $363,000 a year in the Washington area and only 1% to 2% in the Baltimore area do. Only 1% of US workers earn this much, which reduces the chance that the average visitor would find the maglev a prudent way to travel between Baltimore and Washington.

The just-described calculation for an individual traveler is the middle calculation shown schematically in Figure 1.

For a family traveling together, the picture is even less rosy than for an individual traveling alone. With more than one wage earner in many families, household income is often higher than individual income, but a family of four would need four maglev tickets. Few families would think it a good deal to save a few minutes on a trip between Baltimore and Washington by paying $160–$320 for four tickets instead of driving and paying about $7 for gas and car maintenance. As shown in this chapter's appendix, annual household income would have to be more than $1.6 million for a

---

9. $182 h⁻¹ = $53 · 60 min. h⁻¹ ÷ 17.5 min.

family of four to consider the maglev reasonably priced under these conditions. Fewer than 1% of households earn this much in the Baltimore-Washington region.

The calculations in the present chapter have so far assumed that people are willing to pay for travel-time savings at a rate of 100% of their hourly income. The next section varies this assumed value of willingness to pay and shows that the same conclusion can be drawn.

Willingness to Pay

By analyzing many surveys and traffic studies, transportation modelers have found that people are typically willing to pay no more than 50% to 150% of their hourly income to save an hour of travel time.[10]

If one varies the traveler's willingness to pay from 50% to 150%, one arrives at a range of incomes at which the individual would find the maglev ticket price worth the travel time saved. The lower someone's willingness to pay, the higher their income would need to be before the maglev would seem like an attractive proposition. The range of incomes is $242,000 and $727,000, as worked out in the appendix of the present chapter. Figure 1 illustrates this range at the points labeled "low" and "high."

If most Baltimore-Washington residents had a low willingness to pay for travel-time savings, it would result in less than 1% of individual workers finding the maglev attractive. If most of the region's residents had a high willingness to pay for travel-time savings then approximately 4% of them would find the maglev an attractive option.

Under no combination of assumptions would anywhere near a majority of the region's residents earn enough that the maglev's travel-time savings would justify its ticket price.

Serving the 2%, Kind of

In a sense, the 2% figure calculated in the present chapter overstates the market share of the proposed Baltimore-Washington maglev. The unmentioned issue is that the maglev would serve only a small portion of the region. Chapter 3 of the present document shows that the maglev's three stations could save people travel time only on the small fraction of possible trips that start and end fairly close to a maglev station (page 41).

Another approach to estimating the maglev's maximum-possible market share is to consider commuter data from the US Census. The Census Bureau has determined that less than 1% of the region's workers commute between Baltimore and Washington. This statistic means that, even if the maglev could somehow capture all of these commuters, it would still serve only 1% of the workforce.

The maglev would arbitrarily and disproportionately benefit the small fraction of the region's wealthy who make frequent trips between downtown Baltimore and Washington and whose trips begin and end near maglev stations. The rest of the wealthy would be poorly served by the maglev.

In summary, one might say that the maglev would serve 1% of the 2%. The people who would use the maglev would be both rich (2% of the population) and geographically lucky, i.e., part

---

10. The appendix of the present chapter discusses the use of this rule of thumb in the transportation-modeling field.

of the 1% or so of the region's population that frequently travels between the two cities.[11]

Conclusion

The people who would ride the proposed Baltimore-Washington superconducting maglev would be drawn from the wealthiest 2% of the region's population. The calculation that supports this prediction has two steps. In the first step, the concept of "willingness to pay" is used to estimate the income that an individual would need before the maglev would seem like a good deal given the ticket price and travel-time savings. In the second step, the income distribution reported in the US Census is used to determine what percent of the region's population earns this much.

To a first approximation, someone would have to earn at least $363,000 a year before the maglev's travel-time savings would seem worth its ticket price. Only 2% of workers earn this much in the Baltimore-Washington region.

Varying a person's willingness to pay for travel-time savings would result in a range for the minimum income needed to make the maglev an attractive option: an annual income of $242,000 to $727,000. Approximately 4% of workers in the Baltimore-Washington region reach the bottom of this income range and fewer than 1% of workers reach the top.

Broadly speaking, the people who would choose to ride the maglev would be more than mere millionaires. They would be earning another million every few years.

While small, the 2% figure overstates the maglev's market share in one sense. The 2% was calculated from the set of travelers contemplating a particular kind of trip. Specifically, a trip in which the maglev would save them time, door to door, compared to other travel options like driving directly to their destination. But few trips start and end close enough to a maglev station to fit in this category, as discussed in Chapter 3 of the present document (page 41). The maglev does not always save time, especially if you have to drive out of your way to reach the maglev station, wait for the train, and then find a ride from the final maglev station to your actual destination.

The Federal Railroad Administration has mostly avoided the question of what portion of the region's population would make use of the maglev. One would hope that elected officials would want to know if the proposed maglev would serve the region as a whole or if the maglev would only serve a small number of wealthy people who happen to live or work near one of the three maglev stations.

By remaining largely silent on this question, the Federal Railroad Administration has made it easier for maglev proponents to broadcast their message. Both before and after the draft environmental impact statement was published, the website of Baltimore Washington Rapid Rail, the company that wants to build the maglev, has claimed that the maglev would be "highly valued" by "all income segments."

---

11. The 2015 American Commuter Survey (ACS) of the US Census Bureau reported 1.829 million employed people in the following five jurisdictions: District of Columbia, Alexandria, Arlington County, City of Baltimore, and Baltimore County. The ACS also reported that 13,087 of these employed people either worked in Baltimore and lived in Washington or vice versa: US Census Bureau (2014, 2015). These 5 jurisdictions are, to a first approximation, the geographic extent of the maglev ridership area: see Chapter 3 on page 41.

References

AAA, 14 December 2020: *Your Driving Cost: 2020*. 8 pp, https://newsroom.aaa.com/wp-content/uploads/2020/12/2020-Your-Driving-Costs-Brochure-Interactive-FINAL-12-9-20.pdf.

Boardman A. E, D. H. Greenberg, A. R. Vining, and D. L. Weimer, 2018: *Cost-Benefit Analysis: Concepts and Practice*. 5th ed., Cambridge Univ. Press, 594 pp.

EPA, 2019: *The 2019 EPA Automotive Trends Report*. 211 pp, https://nepis.epa.gov/Exe/ZyPDF.cgi?Dockey=P100YVFS.pdf. Page 5 states the average fuel economy of 25.1 miles per gallon for new cars.

Federal Railroad Administration, January 2021: *Baltimore-Washington Superconducting MAGLEV Project Draft Environmental Impact Statement and Draft Section 4(f) Evaluation*. 3,053 pp. (main text and appendices), https://bwmaglev.info/index.php/project-documents/deis.

Gas Buddy, 2021: 120-month average retail price chart. web page, https://www.gasbuddy.com/. States the average price per gallon of gas in the United States, the Washington area, and the Baltimore area.

Khattak, A., A. Kanafani, and E. Le Colletter, 1994: *Stated and Reported Route Diversion Behavior: Implications on the Benefits of ATIS*. research report, Institute of Transportation Studies, Univ. California, Berkeley, ISSN 1055-1425, 36 pp, https://escholarship.org/uc/item/4fz4h20k.

Louis Berger, 8 November 2018: *Baltimore-Washington SCMAGLEV Project Final Ridership Report*. As of July 2021, this document is unavailable to the public except in an extremely redacted form. Page 48 of this document states the 67% figure discussed in the present chapter according to the FRA (2021), maglev DEIS, Chapter 4.6, page 4.6-3, footnote 9.

Meyer, J. R., W. B. Tye, C. Winton, and J. A. Gomez-Ibanez, 1999: *Essays in Transportation Economics and Policy: A Handbook in Honor of John R. Meyer*. Brookings Institute Press, https://play.google.com/books/reader?id=MFhkehz-Ky0C&hl=en&pg=GBS.PA42. Page 42 cites McFadden, Travitie, and Associates (1966, pg. 116).

Ortuzar, J., and L. G., Willumsen, 2011: *Modeling Transportation*. 4th ed., Wiley, 586 pp.

US Census Bureau, 2014: *American Community Survey: Design and Methodology*. 222 pp, https://www2.census.gov/programs-surveys/acs/methodology/design_and_methodology/acs_design_methodology_report_2014.pdf.

US Census Bureau, 2015: Table 4, Residence MCD/County to Workplace MCD/County Commuting Flows for the United States and Puerto Rico Sorted by Workplace Geography: 5-Year ACS, 2011-2015. An Excel spreadsheet for the entire country with over 594,000 rows. On the web page titled "2011–2015 5-year ACS commuting flows," https://www.census.gov/data/tables/2015/demo/metro-micro/commuting-flows-2015.html.

US Dept. of Transportation, 2016: *The Value of Travel Time Savings: Departmental Guidance for Conducting Economic Evaluations Revision 2 (2016 Update)*. 26 pp, https://www.transportation.gov/office-policy/transportation-policy/revised-departmental-guidance-valuation-travel-time-economic.

Whittington, D., and J. Cook, 2019: Valuing changes in time use in low- and middle-income countries. *Journal Benefit-Cost Analysis*, **10**, 51–72, https://doi.org/10.1017/bca.2018.21. Earlier draft: Guidelines for Benefit-Cost Analysis, Working Paper #1, Benefit-Cost Analysis Reference Case Guidance Project.

Willumsen, L., 2014: *Better Traffic and Revenue Forecasting*. Maida Vale Press, 258 pp.

Appendix

This appendix describes Tables 1 and 2 on pages 36 and 37. These tables provide evidence for the conclusions drawn by the present chapter. Table 1 shows the minimum income that a customer would likely have if they found the maglev ticket price worth the limited amount of travel-time savings relative to making the trip by car. Table 2 shows the income distribution for the region's residents. Together, these tables support the conclusion that approximately 2% of the region's workers are wealthy enough to prefer the maglev over driving when contemplating a trip that would start and end close to one of the three proposed stations.

The Income Equation

The following equation calculates the minimum annual income required for someone to find the maglev as good a value as driving directly to the destination given the maglev ticket price and the limited travel-time savings associated with it.

Equation (1)

$I = (\ 2{,}000\ h\ y^{-1}\ )\ c\ (\ 60\ min.\ h^{-1} \div t\ )\ (100\% \div w\ )$

In Equation (1), variable I has units of dollars per year, and c (dollars) is the cost difference between riding the maglev and driving directly to the destination. The amount of travel time saved by riding the maglev is t (minutes). Variable w (percent) is the percentage of hourly income that a customer is willing to pay to save an hour of travel time. The initial factor of 2,000 converts from hourly to annual income based on the round numbers of 40 hours per workweek and 50 workweeks per year.

For an individual traveler, the cost difference c is $33 to $73. This estimate is based on a maglev ticket price of $40 to $80 one-way per person and a $7 cost of making the trip by car ([33,73] = [40,80] - 7).

A $40-to-$80 maglev ticket price is used in the draft environmental impact statement (DEIS) to forecast maglev ridership. In this context, ridership is defined as the number of one-way trips per year made in the year 2030 or 2045.[12]

The $7 cost of driving is calculated as follows. Fuel plus maintenance is $7.08 for a 39.6-mile-long trip between Washington and Baltimore. The fuel-plus-maintenance cost is the trip length multiplied by the AAA estimate of $0.1787 for a typical car's per-mile cost for gas and maintenance (7.08 = 39.6 · 0.1787). Gas alone would cost about $3.94. Gas cost may be calculated based on the trip length, a typical car's fuel economy of 25.1 mile per gallon, and a fuel price of $2.50 per gallon (3.94 = 2.50 · 39.6 ÷ 25.1). The last section of the present appendix discusses alternatives to the $7 estimated driving cost.[13]

The cost difference c in Equation (1) would be greater for a family of four than for an individual traveler because the family would require four one-way maglev tickets or would travel together in a single car. For a family of four, the cost difference would be $153 to $313 (i.e., 4 · [40,80] - 7). For estimating a family's transportation choice when traveling together, it is plausible to consider household income, while for an individual traveling alone, it is plausible to consider individual income. Such assumptions are acceptable when calculating a ballpark estimate to double-check the reasonableness of an official ridership forecast.

---

12. $40–$80 ticket price: Appendix D2, pg. D-107, D-108, "Final SCMAGLEV Fare Assumptions" section.

13. 39.6-mile trip length: Appendix D4, Table D4-59, pg. E-82; miles per gallon: EPA 2019; approximate average dollars per gallon for 2015 to 2021: Gas Buddy 2021; $0.1787/mile for medium sedan: AAA 2020.

In Equation (1), travel time saved t is 8 to 27 minutes, a value stated in the DEIS.[14]

The quantity w in Equation (1) is known as "willingness to pay" in the transportation-modeling field. In this text, w is expressed as the percentage of hourly earnings representing the maximum amount that the customer would be willing to pay to save an hour of travel time.

Willingness to Pay

Willingness to pay is an empirical guideline that the transportation-modeling profession has derived by summarizing many transportation studies.

Various studies recommend values for willingness to pay that are typically between 25% and 140% of hourly earnings. The US Department of Transportation (2016, Table 1, pg. 13) recommends 70% of pre-tax household income for commuters traveling on high-speed rail. Whittington and Cook (2017) recommend 25% to 75% of after tax individual income. Khattak et al. (1993) suggest 50% of income, but state that the percentage becomes much lower (10%) if the new form of transportation is only a slight improvement over the consumer's current form of transportation. Boardman et al. (2018, pg. 393) suggest 40% to 50% of income as the willingness to pay if the travel is enjoyable. Willumsen (2014, pg. 89) suggests 50% to 80% for commuting, 50% to 60% for non-work travel, and 110% to 130% for travel during work hours. Ortuzar and Willumsen (2011, pg. 511) suggest that commuting and non-work travel time is valued at 25% to 43% of the hourly income of individual full-time workers. Meyer et al. (1999) suggest that travelers may be willing to spend more (140% of hourly earnings) for travel during office hours or in heavy traffic.

Willingness to pay is a quantity used to forecast the transportation decision of an individual traveler within a model forecasting the ridership of a transportation project. Willingness to pay should not be confused with another quantity usually called the "equity value of time." Confusion is possible, in part, because both quantities are sometimes referred to by the same acronym, VTTS, which stands for "value of travel time saved."

Equity value of time is a quantity used in cost-benefit analysis to calculate the total benefit to society of a transportation project. Federal regulations stipulate that the same dollar value shall be attached to each person's travel time, regardless of that person's income. For the maglev, the DEIS used \$15.20 h^{-1} or \$27.10 h^{-1} as the equity value of time for personal or business travel, respectively. In this context, personal travel is defined as travel outside of office hours.[15]

Because maglev tickets would be so expensive, the equity value of time would be much lower than the willingness to pay demonstrated by someone choosing to ride the maglev. There is nothing surprising about this occurring, but it is something to keep in mind to avoid confusion. Both quantities are expressed in units of dollars per hour.

Evaluating the Income Equation

Equation (1) is evaluated multiple times to create Table 1. Table 1 shows the lower bound of someone's income if he or she finds that the maglev travel-time savings are worth the ticket

---

14. 8–27 minutes saved travel time: Appendix D4, pg. C-6.
15. See discussion in Chapter 5 of Willumsen (2014) and US DOT (2016); \$15.20/h and \$27.10/h: Appendix D4: pg. D-35.

Table 1. The cost of maglev-related travel-time savings (r, dollars per hour) and the income (I, dollars per year) of someone willing to pay that rate to save travel time. The incomes listed in the rightmost three columns are calculated using Equation (1) on page 34.

	Characteristics of a maglev trip, compared to driving			The traveler's minimum annual income for different values of willingness to pay, w [d]		
	Cost difference, c [b]	Travel time saved, t	Cost per hour, r [c]	w = 50%	w = 100%	w = 150%
	Individual, traveling alone					
worst case [a]	$73	8 minutes	$548 h^{-1}	$2.2 million	$1.1 million	$730,000
middle case	$53	17.5 minutes	$182 h^{-1}	$727,000	$363,000	$242,000
best case [a]	$33	27 minutes	$73 h^{-1}	$293,000	$147,000	98,000
	Family of four, traveling together					
worst case	$313	8 minutes	$2,348 h^{-1}	$9.4 million	$4.7 million	$3.1 million
middle case	$233	17.5 minutes	$799 h^{-1}	$3.2 million	$1.6 million	$1.1 million
best case	$153	27 minutes	$340 h^{-1}	$1.3 million	$680,000	$453,000

[a] Worst case and best case refer to cases when the maglev is least or most attractive to travelers considering its ticket price and the amount of travel time saved.

[b] Excess cost per trip is the maglev price (the individual ticket price multiplied by the number of travelers) minus the $7 per-vehicle cost of driving between Baltimore and Washington, as discussed in the appendix.

[c] Cost per hour of travel time saved is calculated as c (60 min h^{-1} ÷ t), using the values for c and t in the two columns to the left.

[d] Willingness to pay is the maximum amount of money that a traveler would be willing to pay to save travel time, expressed as a percent of the traveler's hourly income. The incomes stated in the rightmost three columns of this table are intended to represent the individual income of an individual traveling alone or the household income of a family traveling together.

price. The table shows how the income cutoff varies with maglev-vs.-driving cost difference c, travel time saved t, and willingness to pay w.

Looking at a middle case for travel-time savings and cost difference, Table 1 shows that the lower bound to individual annual income is $727,000, $363,000, or $242,000 for an individual traveling alone who is willing to pay up to 50%, 100%, or 150% of their hourly income to save an hour of travel time.

Table 2 shows that less than 1% of workers in the Baltimore-Washington region would make the upper income quoted ($727,000/year). Table 2 shows that 2% to 3% of Washington-area residents and 1% to 2% of Baltimore-area residents make the middle income ($363,000/year). The table shows that about 5% and 3% of Washington-area residents and Baltimore-area residents, respectively, make the lower income ($242,000/year). For this reason, the main body of the present chapter states

Table 2. Annual income percentiles for individual workers and households in the Washington Metropolitan Statistical Area (MSA), the Baltimore MSA, and the entire United States.[a]

Percentile	Washington MSA		Baltimore MSA		United States	
	Individual[b]	Household	Individual[b]	Household	Individual[b]	Household
99th	$641,000	$910,000	$440,000	$654,000	$363,000	$531,000
98th	$408,000	$660,000	$266,000	$446,000	$257,000	$387,000
97th	$296,000	$522,000	$220,000	$345,000	$217,000	$329,000
95th	$238,000	$387,000	$178,000	$266,000	$176,000	$270,000
92nd	$197,000	$324,000	$149,000	$237,000	$140,000	$221,000
90th	$175,000	$302,000	$135,000	$229,000	$125,000	$201,000
85th	$149,000	$257,000	$116,000	$192,000	$101,000	$166,000
80th	$131,000	$220,000	$100,000	$169,000	$86,000	$142,000
75th	$113,000	$199,000	$90,000	$146,000	$75,000	$124,000
50th	$64,000	$121,000	$51,000	$88,000	$44,000	$68,000
25th	$30,000	$62,000	$25,000	$43,000	$23,000	$34,000
10th	$11,000	$32,000	$10,000	$20,000	$9,000	$16,000

[a] These statistics were published in 2020, and they represent the income reported for the prior 12 months, i.e., January through December, 2019. The data in the table were obtained from the DQYDJ investment blog (https://dqydj.com/income-by-city/) for the Washington and Baltimore MSAs. The statistics for US households were taken from https://dqydj.com/household-income-percentile-calculator/, and for US individual workers from https://dqydj.com/individual-income-by-year/. The DQYDJ blog obtained the data from the IPUMS-CPS research center (https:/doi.org/10.18128/D030.V8.0). The ultimate source of the data is the US Census Bureau.

[b] Individual means "per worker," not per capita.

an average value of 2% or 4% for the fraction of workers in the entire region whose annual income is over $363,000 or $242,000, respectively.

Now, switch from considering an individual traveling alone to a family of four traveling together. Because the cost of maglev tickets for a family of four is so much greater than the cost of driving, their household income would have to be truly extreme for them to choose to ride the maglev on a family outing. Table 1 shows that the necessary annual household income would be approximately $1.6 million. The US Census shows that less than 1% of Washington-area households earn this much, as shown in Table 2.

US Census Data for Interpreting the Income Equation's Output

The US Census Bureau reports the distribution of individual and household income. Values are reported for the country as a whole and for smaller areas. The two areas used in the present chapter are called metropolitan statistical areas (MSAs). Washington and its surrounding suburbs constitute an MSA that is separate and does not overlap with the MSA that contains the City of Baltimore and its suburbs.

The Washington MSA has an above-average income distribution compared to the rest of the country. The percentiles are shown in Table 2. For example, 1% of Washington-area workers earn at least the amount stated in Table 2 for the 99th percentile of individual income. The income distribution in the Baltimore MSA is also above the national average but not as high as the income distribution in the Washington MSA.

The income percentiles in Table 2 were obtained from the DQYDJ financial blog, https://dqydj.com/. The DQYDJ blog obtained the data from the IPUMS-CPS research center, https:/doi.org/10.18128/D030.V8.0. The IPUMS-CPS research center obtained the data from the US Census Bureau.

The individual income values listed in Table 2 represent pre-tax income per worker rather than per capita. The stated value includes income from all sources including wages, investments, and government programs. Workers are included if they are at least 16 years old. The data were published in 2020, and they refer to income earned during the previous calendar year, i.e., the 12 months from January through December, 2019.

Exaggerating the Cost of Driving would not Make Much Difference

While the present chapter uses $60 for the maglev ticket price and $7 for the cost of driving between Baltimore and Washington, other sources suggest a lower maglev ticket price and a higher estimate of the driving cost. These two possibilities do not alter the conclusions of this chapter.

Various hints that maglev tickets would occasionally sell for $27 are irrelevant to the maglev's official ridership forecast. The maglev's draft environmental impact statement (DEIS) states explicitly that it used a $40-to-$80 ticket price to calculate the ridership forecast.[16]

It is worth investigating the possibility that the average consumer may perceive a cost greater than $7 for making a one-way car trip between Baltimore and Washington. The DEIS appears to use a cost of $16.24 for this car trip when calculating the maglev's official ridership forecast. The DEIS, however, is vague on this point. The Northeast Maglev is clear about its estimate of the cost of the car trip: $20.38. The Northeast Maglev is the parent company of the company that wants to build the maglev.[17]

These arguments miss the point. It does not matter that the Federal Railroad Administration and The Northeast Maglev can find ways to assign a high cost to driving between the two cities. What matters is whatever cost the consumer perceives for the car trip when choosing among the available transportation options. Let's be realistic: driving between Baltimore and Washington costs just a few dollars for gas plus perhaps a few dollars set aside for future car maintenance. The trip is so short that the needle of your car's gas gauge barely moves. Depending on whether your destination is downtown or in the suburbs, you might have to pay for parking if you drove there directly, but then again, you'd likely have to pay for parking

---

16. $27 ticket price considered and rejected for $40–$80 ticket price: Appendix D2, pg. D-107, D-108, "Final SCMAGLEV Fare Assumptions" section.

17. $16.24 for 39.6 miles or 0.41 per mile: DEIS, Appendix D4, Table D4-82; pg. D-32; $20.38 for 39.6 miles or 0.56 per mile: The Northeast Maglev Website (24 Apr 2021).

if you drove to one of the three proposed maglev stations.[18]

Whether the consumer perceives that the cost of driving is closer to $7 or $16.24, it would not change the fact that riding the maglev would be financially attractive to only a small number of wealthy people. For example, one can evaluate Equation (1) using $16.24 for the cost of driving and assuming that 83% of hourly income is what the customer is willing to pay for travel-time savings. The result of this calculation is the same as evaluating Equation (1) using a $7 driving cost and a willingness to pay 100% of hourly income. Both 83% and 100% are within the reasonable range of values for willingness to pay, as discussed earlier in this appendix.[19]

---

18. A few dollars for gas: appendix of the present chapter.

19. Use $16.24 driving cost and 83% willingness to pay in Eq. (1) of the appendix of the present chapter to obtain $357,000 = 2,000 h y^{-1} · {$60 - $16.24} · (60 min. hr^{-1} ÷ 17.5 min) · (100% ÷ 83%). This is essentially the same value as the $363,000 when a $7 driving cost and 100% willingness to pay was used earlier in the appendix.

Automobile travel differs from air or rail travel in that it generally involves door-to-door service, offers greater flexibility in time of departure, and does not require travelers to share space with strangers. Consequently, rail travel must be extremely competitive in other dimensions, such as speed or cost, to attract automobile travelers.

—Federal Railroad Administration (2008, pg. 6-7)

3 • GEOGRAPHIC AREA

The proposed Baltimore-Washington maglev would serve a small geographic area

Most maglev customers would start and end their trips near a maglev station, but the project's draft environmental impact statement is vague on this point

A superconducting magnetic-levitation rail line has been proposed to connect Baltimore and Washington. It is important to know where most "maglev" customers would start and end their trips because much of the economic benefit from the maglev may occur in the same area. The maglev's draft environmental impact statement (DEIS) is vague on these topics rather than quantifying benefits by zip code or at least by county.

The DEIS lumps together the maglev's economic impact over a vast area, the Washington-Baltimore Combined Statistical Area defined by the US Census Bureau. In contrast, the DEIS calculates its forecast for the maglev's ridership over a smaller, but still large, area defined by a 25-mile radius around each maglev station. These areas are shown in Figure 1 on the next page.[1]

As explored in this chapter, even a 25-mile radius seems like an overestimate of the maglev's reach. Instead, most maglev customers would probably start and end their trips within a small subset of the 25-mile-radius area around each maglev station. The present chapter suggests that most counties in the Washington-Baltimore Combined Statistical Area would have few if any maglev customers starting or ending their trips there.

Elected officials and the public would like to know which counties and cities would benefit from the maglev and which would be harmed by its construction and operation. For example, the DEIS estimates that 390 to 440 jobs would be created directly or indirectly as a result of operating the maglev, but the DEIS is silent on the question of where these jobs would be located. The maglev's

---

1. CSA: Chapter 4.6, pg. 4.6-1; 25-mile radius: Appendix D2, pg. C-106.

Figure 1. A map showing four areas discussed in relation to the proposed Baltimore-Washington maglev. First, the thick line indicates the outer boundary of the Washington-Baltimore Combined Statistical Area. Second, counties outlined in white are in the jurisdiction of two planning bodies: the Metropolitan Washington Council of Governments and the Baltimore Metropolitan Council. Third, the dotted gray lines indicate a 25-mile radius from the maglev stations that are proposed at Washington's Mount Vernon Square and Baltimore's Camden Yards. These three areas are mentioned in the maglev's draft environmental impact statement (DEIS). In contrast, the much smaller, dark-gray area is the result of the analysis described in the present chapter. The dark-gray area represents one realization of where most maglev customers would start and end their trips during rush hour. The same dark-gray area is also shown, at a higher magnification, in Figure 2 on page 45.

many negative impacts are quantified in various sections and appendices of the DEIS.[2]

It would have been helpful if the DEIS had plotted contours on a map or used another means to visualize where most maglev customers would start and end their trips. The public cannot find this information in official sources, such as the studies, memos, and data requests that are the source of the DEIS's maglev ridership forecast. These documents are hidden from public view. In fact, their existence is known only from footnotes in the DEIS. Fortunately, enough information is published in the DEIS to guide the analysis in the present chapter.

Background

In January 2021, the Federal Railroad Administration published the draft environmental impact statement (DEIS) for the proposed Baltimore-Washington maglev. The DEIS states that using the maglev would save a traveler 8 to 27 minutes relative to the time the traveler would otherwise spend driving directly to his or her destination.[3]

Because maglev tickets would be so expensive, it is plausible that people would ride the maglev only if it saved them at least 8 to 27 minutes. The DEIS states a ticket price of $40 to $80 per person, one way. The cost of driving between Baltimore and Washington is approximately $7 per car, one way, based on the average trip length stated in the DEIS and the AAA estimate for the cost of fuel and maintenance for a typical car. As a result, the maglev-vs.-car price difference is $33 to $73, one way, with one person in the car, and much more than $33 to $73 with multiple people in the car, such as on a date or family outing.[4]

Travel time saved and travel cost are factors that transportation planners consider when forecasting the ridership for a transportation proposal. The DEIS states that these factors were included in the model that forecasts the ridership for the proposed Baltimore-Washington maglev.[5]

Method

The present chapter identifies the maglev ridership area by exploring where the maglev would save a customer approximately 8 to 27 minutes relative to the amount of time the customer would otherwise have spent driving directly to the destination.

To estimate travel time saved, first pick a trip origin and destination with one point in the Washington area and the other in the Baltimore area. Calculate the time to drive between these two

---

2. 390-440 jobs: Chapter 4.6, pg. 4.6-8. Negative impacts would occur to historical sites (Chapter 4.8); scenic resources (Chap. 4.9); recreational facilities (Chap. 4.7); environmental justice (Chap. 4.5); quality-of-life (Chap. 4.4); hazardous waste sites (Chap. 4.15); forests, forest-interior species, and habitats of rare, threatened, and endangered species (Chap. 4.12); wetlands (Chap. 4.11); economic harm during construction (Appendix D4, pg. D-18 to D-30); and lost revenue for Amtrak and MARC commuter trains (Appendix D4, Table D4-47, pg. D-54).

3. DEIS: FRA 2021; 8-27 minutes: Appendix D4, pg. C-6.

4. The DEIS considered and rejected a maglev ticket price as low as $27 and chose instead to base its ridership forecast on a $40–$80 ticket price: Appendix D2, pg. D-107, D-108; $7.08 cost of making a typical trip between Baltimore and Washington by car based on a 39.6-mile trip length (Appendix D4, Table D4-59, pg. E-82) and a $0.1787-per-mile cost for a medium sedan (AAA 2020).

5. Willumsen 2014, Chapter 5; Ortuzar and Willumsen 2011, Section 15.4; Ridership forecast model: Appendix D2, pg. B-104 to E-110.

points. From that time, subtract how long it would take to travel between the same two points using the maglev. In its simplest form, the maglev trip would include driving to a maglev station, riding the maglev, and riding a car to the destination. Various online applications can provide car travel time between any two points, and time spent on the maglev itself can be estimated from information in the DEIS and other documents. The details of this calculation are described in Appendices 1 and 2 of the present chapter.

The present chapter estimates travel-time savings for a large set of trip origin-and-destination pairs in order to the map where the maglev travel-time savings would be in the 8-to-27-minute range that the DEIS states. The computations are slightly more complicated because two stations are proposed at the Baltimore end of the trip, one at Camden Yards and one at Baltimore/Washington International (BWI) Thurgood Marshall Airport. The solution is to calculate travel time saved for both Baltimore maglev stations, and use whichever value is greater.

One simplification employed in the present chapter is to assume that travel to and from the maglev stations occurs by car, without modeling the option of subway travel to and from the Washington maglev station. Supporting this simplification, the analysis in Appendix 3 finds that, in almost every case, a subway ride would not save time over driving to the downtown Washington maglev station. The existence of the Washington subway has little impact on the geographic extent of the maglev ridership area.

The calculation method is kept simple because the goal is merely to determine whether the maglev ridership area would fill the entire 25-mile-radius area that is studied in the DEIS or if the ridership area would be much smaller than that.

Results

The three sections below identify jurisdictions where most maglev travelers would start and end their trips during rush hour or in light traffic. Also identified are jurisdictions with little or no area served by the maglev regardless of the amount of road congestion. One finding is that the proposed Baltimore-Washington maglev would have an easier time competing against car travel during rush hour than when road traffic is light. In other words, the maglev ridership area is larger during rush hour than when road traffic is light.

The Maglev Ridership Area during Rush Hour

How far one can travel from a maglev station and still save 8 to 27 minutes depends on how close the other end of the trip is to the other maglev station. The figures on the next page show two possible realizations of the maglev ridership area during rush hour. Figures 2 and 3 emphasize access to Washington and Baltimore, respectively.

Figure 2 emphasizes locations at the Washington end of a rush-hour trip while still reaching an appreciable number of locations in Baltimore. Optimized in this way, the maglev ridership area would include about half of the District of Columbia; most of the City of Alexandria, Arlington County, and City of Baltimore; and less than half of the Baltimore County suburbs.

In contrast, Figure 3 shows the maglev ridership area optimized in the opposite way. Figure 3 emphasizes locations at the Baltimore end of the trip. In this case, a portion of eastern Carroll County and northern Anne Arundel County can be reached. This portion of Carroll County is sparsely populated, and this part of northern Anne Arundel County contains Glen Burnie and Pasadena. Few people would use the maglev in this scenario because most of the

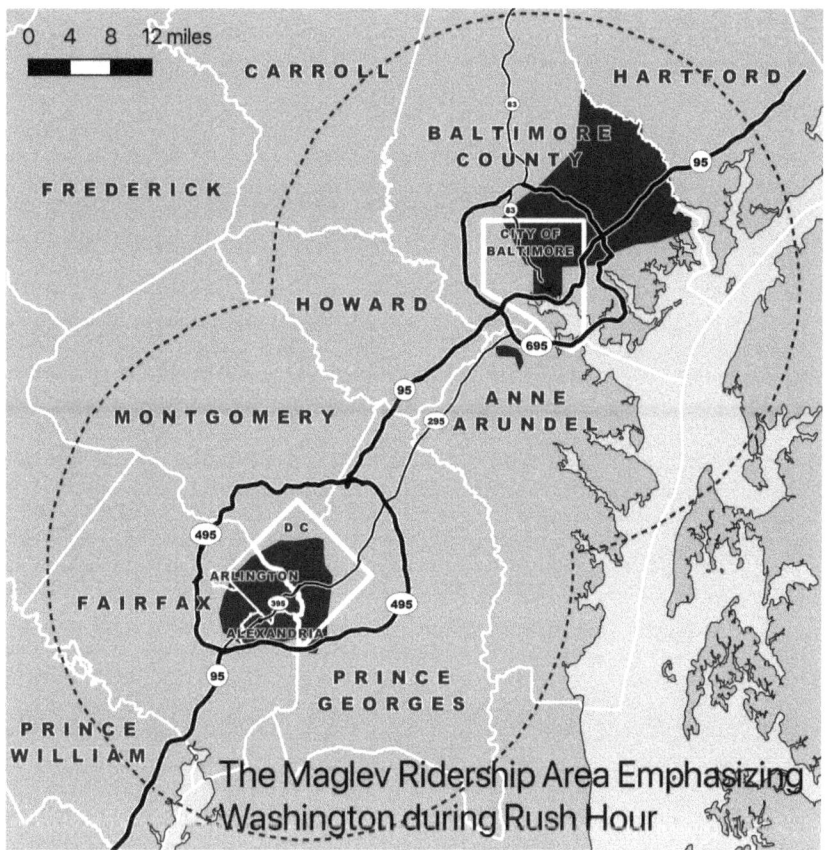

Figure 2. During rush hour, the maglev ridership area (shown in red). This area is optimized to reach many locations at the Washington end of the trip. The ridership area is reduced at the Baltimore end of the trip so that the goal can still be realized of the maglev trip saving the traveler at least 8 minutes of travel-time relative to the time that would otherwise be spent driving directly to the destination.

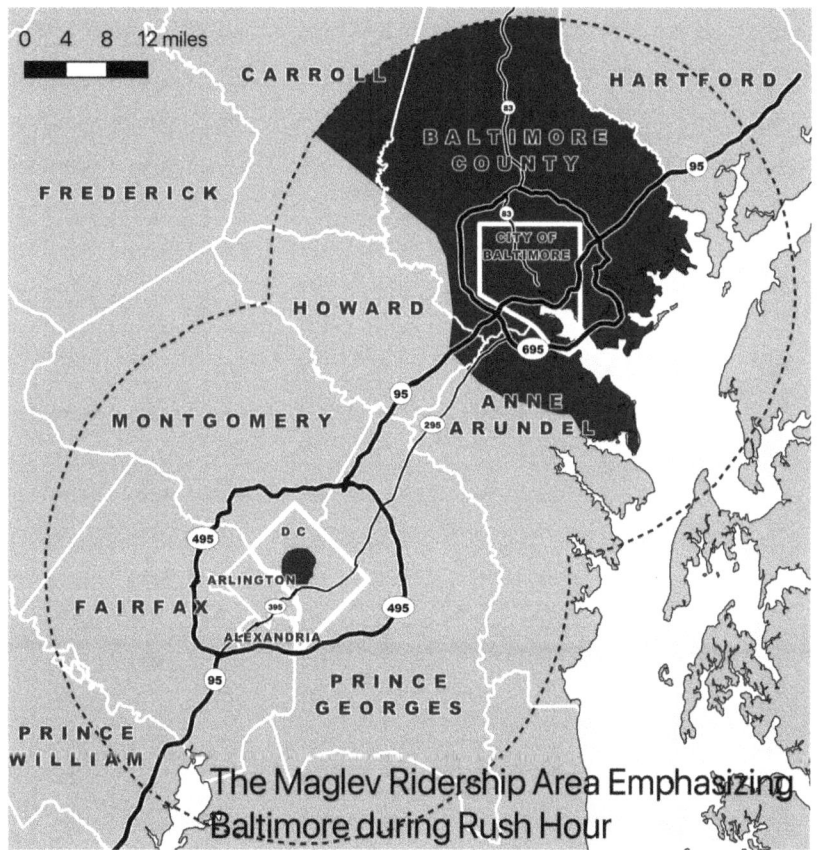

Figure 3. The same as Figure 2 except that the maglev ridership area (shown in red) is optimized to reach more locations at the Baltimore end of the trip during rush hour. Simultaneously, locations reachable at the Washington end of the trip are reduced so that the goal can still be realized of the maglev trip saving the traveler least 8 minutes of travel time relative to the time to drive directly to the destination. Few people would make use of the maglev under these circumstances because few locations can be reached in Washington.

District of Columbia can be reached including Capitol Hill, most residential area, and the federal offices just south of the National Mall.

The DEIS forecasts that approximately 15% of maglev travelers would be airline passengers headed to or from BWI airport, and the present chapter neither confirms nor questions this percentage.[6]

The present chapter does, however, suggest that the BWI maglev station would have limited utility for points other than the airport's main terminal. Figure 2 shows a small maglev ridership area located to the north and east of BWI. From the rest of the business parks and residential areas within a few miles of BWI, one can easily reach Interstates 95 and 295, which are direct routes to Washington. The BWI main terminal can be reached from only one direction (west), and the airport's main road loop can be slow due to congestion. In this way, the existing road network would geographically isolate a maglev station adjacent to the BWI main terminal.

To summarize the rush-hour results, the maglev would save the traveler approximately 8 to 27 minutes in an area much smaller than the DEIS-supplied 25-mile radius around the maglev stations.

Careful examination of Figures 2 and 3 reveals the maglev ridership area is bunched to the side of the maglev station furthest from the other maglev station. In other words, the maglev ridership area is mostly south and west of downtown Washington and north and east of downtown Baltimore. This makes sense because using the maglev won't save much time on trips that start and end between the two cities. In this case, traveling to and from the maglev station would take you far out of your way.

Another factor evident in Figures 2 and 3 is that the maglev ridership area is larger at the Baltimore end of the trip than at the Washington end. This asymmetry is due to the fact that the proposed Washington maglev station at Mount Vernon Square would be in the middle of an area with especially slow rush-hour traffic and many traffic lights. In contrast, the maglev station proposed for Baltimore's Camden Yards would be a short detour from routes that would take drivers initially south and west toward downtown Baltimore along Route 83 and Interstate 95 and subsequently south toward Washington.

The present chapter excludes Hartford County from the maglev ridership area at the Baltimore end of the trip. Slightly less than half of Hartford County is within the DEIS's 25-mile radius from the maglev station proposed at Baltimore's Camden Yards. The lack of an existing market for the maglev in Hartford County is indicated by US Census data. The Census data show that almost no Hartford County residents commute to jobs in Washington and few Washington residents commute to jobs in Hartford County. Following the same sort of logic, the DEIS states that it shrunk or expanded its 25-mile-radius area, as necessary, to reflect existing travel patterns.[7]

The Maglev Ridership Area when Road Traffic is Light

When road traffic is light, only a small portion of downtown Washington and downtown Baltimore would be included in the ridership area shown in Figure 4. Two factors contribute to the smallness of the ridership area in light traffic. Directly

---

6. Only 14.5% of maglev trips would be downtown-to-airport, with the remaining 85.4% downtown-to-downtown: Appendix D4, Table D4-25, pg. D-42.

7. 25-mile radius: Appendix D2, pg. C-106; US Census Bureau (2015).

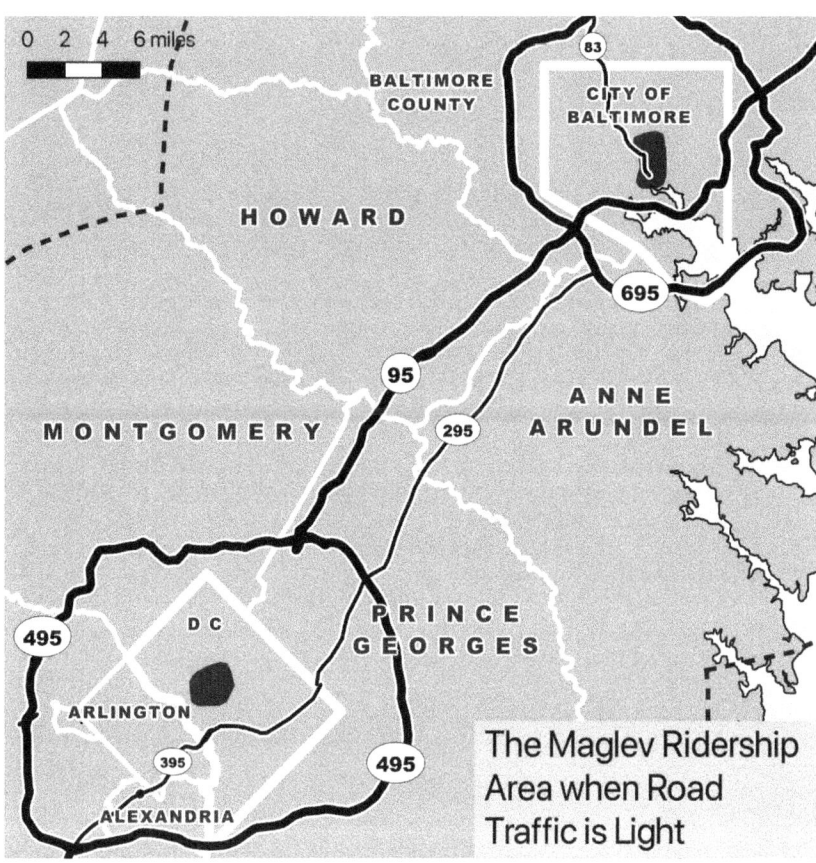

Figure 4. The maglev ridership area when road traffic is light. Because road travel is so much faster in light traffic and maglev trains would be less frequent outside of rush hour, the maglev can outcompete car travel over a smaller area in light traffic than during rush hour. About 10 to 18 minutes of travel time would be saved when road traffic is light if both the trip origin and destination are in the portion of downtown Baltimore and downtown Washington that are colored dark gray in this map.

driving to the destination is much faster in light traffic than during rush hour. In addition, maglev trains would be less frequent outside of rush hour, and therefore one would wait longer for the next train.

When road traffic is light, the utility of the maglev is limited in two additional ways. Appendix 2 of the present study suggests that the maglev would save travelers only 10.5 to 17.5 minutes of travel time when road traffic is light, i.e., the lower half of the target range of 8 to 27 minutes. Such limited travel-time savings suggests that only wealthier travelers would find the maglev travel-time savings sufficient to justify the $40-to-$80 maglev ticket price outside of rush hour.

The non-rush-hour utility of the maglev is also limited because the ridership area depicted in Figure 4 applies only to various non-rush-hour times during which there are maglev train departures at least every 15 minutes. In contrast, maglev train departures that are 30 minutes apart may occur during off-peak weekend hours, and at these times, the maglev ridership area would essentially disappear.

Jurisdictions not Served by the Maglev

The present chapter finds that many counties are outside of the area served by the maglev but are included in the DEIS study area. Based on the travel-time analysis in the present chapter, elected officials and members of the public should read with skepticism any claim that the Baltimore-Washington region, as a whole, would benefit from the maglev rather than a few small areas near a maglev station.

Even during rush hour, few if any maglev customers would start or end their trip in the majority of the counties within the jurisdiction of the Metropolitan Washington Council of

Governments or the Baltimore Metropolitan Council. These counties are outlined in white in Figure 1 (page 42). In addition, few maglev customers would start or end their trips in most counties in the Washington-Baltimore Combined Statistical Area, which is also shown in Figure 1. In fact, few if any maglev customers would even pass through most of these counties on their way to or from a maglev station.[8]

Conclusion

The present analysis compares travel time between Baltimore and Washington when a trip is made using the proposed superconducting maglev or entirely by car. The results of the analysis are maps of the maglev ridership area, the area near maglev stations where most maglev customers would start or end their trip.

In this chapter, the maglev ridership area is modeled as the area where the maglev would save a traveler approximately 8 to 27 minutes compared to the time that the traveler would otherwise spend driving directly to his or her destination. The maglev's draft environmental impact statement (DEIS) asserts that a maglev customer would save this much time. More importantly, travel-time savings this great are a plausible prerequisite for people who travel between Baltimore and Washington to find the maglev an attractive option considering its one-way $40-to-$80 ticket price per person.

During rush hour, the present chapter finds that the maglev would save travelers about 8 to 27 minutes on trips that start and end in at least half of the area of each of these jurisdictions: the District of Columbia, the City of Alexandria, Arlington County, Baltimore County suburbs, and the City of Baltimore. Even during rush hour, few if any maglev customers would start or end their trips in the majority of the counties within the jurisdiction of the Metropolitan Washington Council of Governments or the Baltimore Metropolitan Council.

When road traffic is light, the maglev ridership area would be even smaller. It would include, at most, only a portion of downtown Washington and downtown Baltimore. The reason for the maglev's limited utility when road traffic is light is that there would be fewer maglev trains per hour than during rush hour and car travel would be much faster.

References

AAA, 14 December 2020: *Your Driving Cost: 2020*. 8 pp, https://newsroom.aaa.com/wp-content/uploads/2020/12/2020-Your-Driving-Costs-Brochure-Interactive-FINAL-12-9-20.pdf.

Baltimore Metropolitan Council (BMC), 2020: *Maximize2045: A Performance-Based Transportation Plan*. https://www.baltometro.org/transportation/plans/long-range-transportation-plan/maximize2045.

Department of Transportation (DOT), 2011: *High-speed/Intercity Passenger Rail (HSIPR) Best Practices: Ridership and Revenue Forecasting*. 89 pp, https://www.oig.dot.gov/sites/default/files/files/OIG-HSR-Best-Practice-Ridership-and-Revenue-Report.pdf.

Department of Transportation (DOT), 1997: *Transfer Penalties in Urban Choice Modeling*. published by the Travel Model Improvement Program, DOT-T-97-18, 51 pp, https://babel.hathitrust.org/cgi/pt?id=ien.35556028271757.

Encyclopedia Britannica, 2020: Escalator. https://www.britannica.com/technology/escalator.

Federal Railroad Administration, January 2021: *Baltimore-Washington Superconducting MAGLEV Project Draft Environmental Impact Statement and Draft Section 4(f) Evaluation*. 3,053 pp. (main text

---

8. Map of counties in MWCOG (2010, pg. 5) and BMC (2020, pg. 6).

and appendices), https://bwmaglev.info/index.php/project-documents/deis.

Flynn, W. J., 6 May 2021: Testimony before the US House of Representatives Subcommittee on Railroads, Pipelines, and Hazardous Materials. Hearing title: *The Benefits and Challenges of High-Speed Rail and Emerging Rail Technologies*, 20 pg., https://transportation.house.gov/imo/media/doc/Flynn%20Testimony2.pdf.

Guo, Z., and N. H. M. Wilson, 2004: Assessment of the Transfer Penalty for Transit Trips. *J. Transportation Research Board*, No. 1872, 10–18, https://journals.sagepub.com/doi/pdf/10.3141/1872-02.

Metropolitan Washington Council of Governments (MWCOG), 2010: *Region Forward Vision*. 72 pp, https://www.mwcog.org/regionforward/.

Microsoft, 2018: Bing Maps Route API, https://docs.microsoft.com/en-us/bingmaps/rest-services/routes/.

Ortuzar, J. de D., and L. G. Willumsen, 2011: *Modeling Transportation*. 4th ed., Wiley, 586 pp.

Roy, S., 7 December 2017: How many people commute between Baltimore and DC? *Greater Greater Washington* blog, https://ggwash.org/view/65822.

Titus, J., 2 February 2015: Governor Hogan thinks only 10% of Marylanders use transit. Actually, 25% or more do. *Greater Greater Washington* blog, https://ggwash.org/view/36978.

US Census Bureau, 2015: Table 4, Residence MCD/County to Workplace MCD/County Commuting Flows for the US and Puerto Rico Sorted by Workplace Geography. 2011–2015 5-Year American Community Survey (ACS) Commuting Flows, https://www.census.gov/data/tables/2015/demo/metro-micro/commuting-flows-2015.html.

US Census Bureau, 2017: *American Community Survey Information Guide*. ACS-331(C)(2017), 18 pp, https://www.census.gov/content/dam/Census/programs-surveys/acs/about/ACS_Information_Guide.pdf.

Willen, C., K. Lehmann, and K. S. Sunnerhagen, 2013: Walking Speed Indoors and Outdoors in Healthy Persons and in Persons with Late Effects of Polio. *J. Neurology Research*, **3**, 62–67, https://doi.org/10.4021/jnrl187w.

Willumsen, L., 2014: *Better Traffic and Revenue Forecasting*. Maida Vale Press, 258 pp.

List of Appendices

This chapter contains three appendices. The first describes a method for generating travel-time penalty maps. Appendix 2 explains how these penalty maps can be used to estimate the maglev ridership area. Appendix 3 validates one of the approximations that are built into these maps, namely that it is acceptable to model car travel to and from the Washington maglev station rather than using the subway to travel to and from the maglev station. These appendices begin on the following pages:

Appendix 1	page 50
Appendix 2	page 59
Appendix 3	page 62

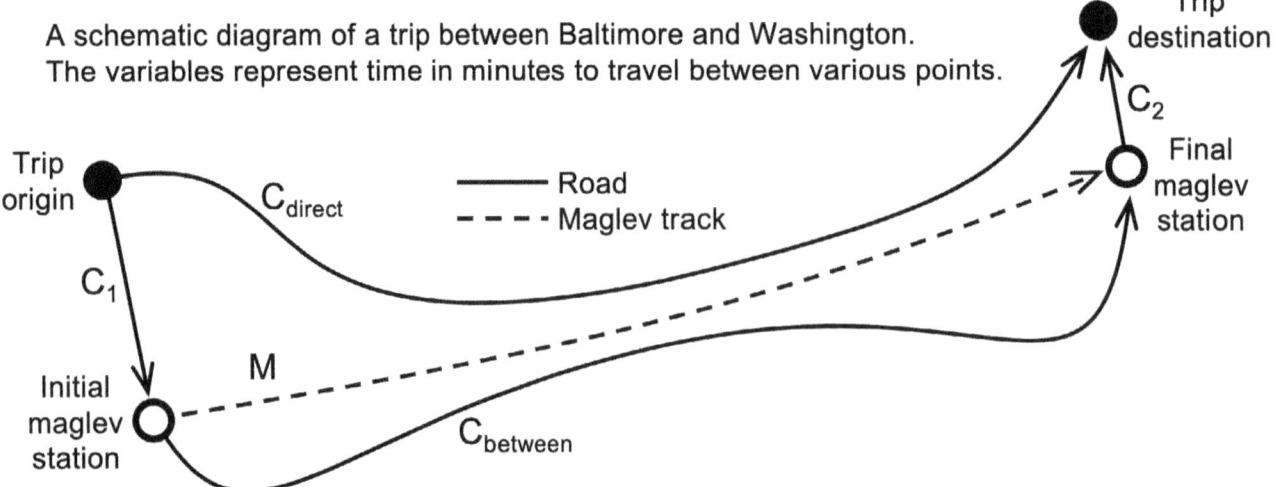

Figure 5. A schematic diagram of a maglev-assisted trip. Also shown is the trip made by driving directly to the destination, which takes Cdirect minutes to complete. The time to drive from one maglev station to the other is Cbetween. The diagram shows the trip in one direction only. Appendix 1 of the present chapter averages the time to make the trip in both directions.

Appendix 1

Travel-Time Penalty Maps that Identify Travel Time Saved Relative to the Maximum Travel Time Saved

A travel-time penalty map shows the geographic variation of travel time saved. That is, a penalty map shows travel time saved as a function of how far a trip terminus (origin or destination) is from the nearest maglev station.

To calculate the travel-time penalty, first calculate the time it would take to drive directly to the destination and the total travel-time to take a maglev-assisted trip, a trip in which the customer drives to and from maglev stations and rides the maglev itself.

The Components of Total Travel Time for a Maglev-Assisted Trip

To estimate total travel time, one must first define the components of total travel time of a maglev-assisted trip. This appendix assumes travel to and from a maglev station would occur by car, a reasonable simplification for two reasons. Most people use cars rather than public transit to travel to work or other destinations. Furthermore, Appendix 3 of the present chapter shows that taking the subway to the maglev station would be no faster than driving to the maglev station in most cases (page 62).[9]

The total travel time of a maglev-assisted trip may also be called door-to-door time, as is done in

---

9. 9% of Maryland workers use public transit to travel to work and 25% use public transit overall: Titus, 2015.

the maglev's draft environmental impact statement (DEIS). Chapter 5 of the present document defines the term "maglev-assisted trip" to emphasize that a maglev ride would be only one leg of a maglev customer's trip (page 85). In this chapter, the total travel time of a maglev-assisted trip is modeled as the sum of the terms in Equation (1). All of these terms have units of minutes and are described in subsequent paragraphs. Some of these terms are shown in Figure 5.[10]

(1) *Total duration of a maglev-assisted trip*
$T_{maglev} = C_1 + W + T + TP + LH + C_2$

In Equation (1), C_1 is the time it takes to travel by car from the trip origin to the nearest maglev station and C_2 is the time required to ride a car from the final station to the destination. These car-travel times are estimated by calling the Route Application Programming Interface (API) of Microsoft's Bing Maps, as described subsequently.

In Equation (1), LH is the line-haul time, which means the time spent on the maglev while it travels between the initial and final stations. The DEIS states that the maglev's line-haul time would be 15 minutes between the stations proposed at Washington's Mount Vernon Square and Baltimore's Camden Yards. The DEIS does not state a numerical value for the typical door-to-door duration of a maglev-assisted trip. The DEIS also does not mention that the line-haul time for Amtrak's Acela train service between Baltimore and Washington would be 21 minutes after planned track improvements—only slightly longer than the proposed maglev's line-haul time.[11]

In Equation (1), W is the average wait time for the next maglev train, which is simulated as half the time between scheduled departures of the maglev. The time between departures is called "head time" by transportation modelers.

Based on the scheduling information published in the DEIS, three values for average wait time W are considered in the present chapter: 4, 7.5, and 15 minutes. The wait time would average 4 minutes during weekday rush hour. The wait time would average 7.5 minutes outside of rush hour on weekdays and during peak weekend travel times. Last, the wait time would average 15 minutes during off-peak times on weekends.

These three values for average wait time W are based on the weekday schedule published in the DEIS. This schedule shows maglev departures every 8 to 15 minutes in each direction. The DEIS notes that, on Saturday or Sunday, there would be half as many trains as on a weekday, so at some times, the maglev trains would likely depart only every 30 minutes in each direction.[12]

In Equation (1), T is the transfer time of 6 minutes. Transfer time is the sum of time that the customer would spend walking or riding escalators inside of the initial and final maglev stations. Based on the station diagrams in the DEIS, the time used to travel from the car drop-off location to the train platform, or vice versa, would be approximately 3 minutes at each end of the trip. This 3-minute per-station estimate is based on a 100-meter walking distance plus a 35-meter vertical ascent or descent by escalator. Duration is calculated as distance divided by speed, and walking speed is about

---

10. Door-to-door: Appendix D4, pg. D-36.

11. 15-minute maglev "travel time": Chapter 4.2, Table 4.2-1, pg. 4.2-5, also in Appendix D4, footnote to Table D4-59, pg. E-82. 21 minutes on Amtrak Acela: Flynn (2021).

12. 15-minute maglev "travel time": Chapter 4.2, Table 4.2-1, pg. 4.2-5, also in Appendix D4, footnote to Table D4-59, pg. E-82.

1.54 m s^{-1}. An escalator travels slantwise at about 36 meters per minute at a 30° slope. Time spent walking and riding escalators may be calculated as follows: 3 min ≈ 100 m ÷ (1.54 m s^{-1} · 60 s min^{-1}) + 35 m ÷ (36 m min^{-1} · sin 30°).[13]

The DEIS acknowledges that maglev wait time W and transfer time T are components of total travel time, but the DEIS does not reveal the value or range of values that it uses for these two quantities. Worse yet, the DEIS makes no mention of another component of total travel time, a component that transportation modelers call "transfer penalty."[14]

The transfer penalty for switching between car and maglev during a single trip is represented by TP in Equation (1). It is common practice for transportation models to include a transfer penalty that represents the disincentive that customers feel when they contemplate a trip that would require multiple modes of transportation to complete. Transfer penalty is expressed in minutes, and it is an additional factor beyond the actual minutes that the customer would spend to perform the mode transfer. This chapter uses a transfer penalty TP of 4 minutes in addition to the time T required to walk and ride escalators within maglev stations. Four minutes is a conservative estimate for the transfer penalty because the US Department of Transportation recommends a transfer penalty of 12 to 15 minutes for a trip that includes one or more mode transfers.[15]

Having an expression for maglev total travel time (T_{maglev}, Equation 1, page 51) makes it possible to calculate the travel time saved by taking a maglev-assisted trip instead of driving directly to the destination (ΔT, Equation 2):

(2) *Travel time saved*

$\Delta T = C_{direct} - T_{maglev}$

$\Delta T = C_{direct} - (C_1 + W + T + TP + LH + C_2)$

In Equation (2), the time to drive directly to the destination is represented by C_{direct}.

The Maximum Travel Time Saved

The maximum travel-time savings would occur in a small category of origin-destination pairs. In this category, both maglev stations are located along an optimal route that a car traveler could drive from trip origin to trip destination. In other words, the maglev would save a traveler the most time if stopping at a maglev station would not take the traveler out of his or her way.

If both maglev stations are on the way when driving between trip origin and destination, then the time to drive directly to the destination C_{direct} can be expressed as $C_1 + C_{between} + C_2$ where $C_{between}$ is the time to drive between the two maglev stations (Figure 5, page 50). In this theoretical situation, the C_1 and C_2 terms can be canceled from the general equation for travel time saved (Eq. 2) to arrive at a shorter expression (Eq. 3) that represents the theoretical maximum travel

---

13. 100 m horizontally by walking and 35 m vertically by escalator: Appendix D4, pg. D-36; 30° slope and 36 m min^{-1} escalator speed: Encyclo. Britannica 2021; 1.54 m s^{-1} walking speed: Willen et al. 2013, pg. 66.

14. "Transfer and wait times out-of-vehicle as well as in-vehicle time": Appendix D4, pg. D-36; transfer penalty important: Willumsen 2014, pg. 210.

15. Transfer penalty equivalent to 12–15 minutes of in-vehicle time: DOT 1997, pg. 40; similar values for transfer penalty: Guo and Wilson 2004, Table 1; use the same value for transfer penalty if the trip includes 1 or more mode changes: DOT 1997, pg. 41; transfer penalty is "typically expressed in terms of time" and "is over and above any actual travel or connecting time, as transfers are often the most onerous aspect of a trip": DOT 2011, pg. 47.

time saved. No trip origin-destination pair appears to achieve this theoretical maximum value. A few origin-destination pairs were found with travel-time savings that were a few minutes less than this theoretical maximum.

(3) *Maximum travel time saved*

$$\Delta T_{max} = C_{between} - (W + T + TP + LH)$$
$$\Delta T_{max} = C_{between} - M$$

where M equals W + T + TP + LH

In Equation (3), the time $C_{between}$ to drive between Mount Vernon Square and Camden Yards is typically about 70 minutes in rush-hour traffic if one averages the duration of northbound and southbound trips. The same trip takes only about 50 minutes in light traffic. The data considered in choosing the round numbers of 50 and 70 minutes are shown in Table 1. Throughout this chapter, the reported car travel time between two points is the average of the trip duration were it made in either direction. As stated in Table 1, the present chapter examines rush hour and light traffic using trips that begin at 5:00 p.m. Monday and 8:00 a.m. Sunday, respectively.

Equation (3) should be evaluated separately for different trip configurations. For example, the three rows of Table 2 evaluate Equation (3) for rush-hour service, for maximum maglev service during times of light road traffic, and for minimum maglev service during times of light road traffic. Table 2 shows the values of maximum travel time saved, ΔT_{max}, and the associated values for $C_{between}$ and M. The quantity $C_{between}$ depends on the maglev stations used and the level of road congestion. The quantity M depends on the maglev stations used and the frequency of maglev trains.

Creating Travel-Time Penalty Maps

In order to generate a map how far from each maglev station the maglev ridership area extends, it is helpful first to map the degree to which travel to a maglev station takes the customer out of their way. This preparatory map may be called a travel-time penalty map.

The first step to creating a travel-time penalty map is to examine a set of points scattered throughout one city, holding the other end of the trip fixed in the other city. For example, Equation (4) is an expression for the travel time saved when the trip origin is moved all around the city that contains the initial maglev station (so $C_1 > 0$) while the trip destination is fixed at the final maglev station (so $C_2 = 0$):

(4) *Travel time saved when the trip origin is a variable distance away from maglev station 1 and the destination is maglev station 2*

$$\Delta T_1 = C_{direct} - (C_1 + W + T + TP + LH)$$
$$\Delta T_1 = C_{direct} - (C_1 + M)$$

The second step is to calculate the travel-time penalty P_1, which means the degree to which travel to maglev station 1 takes the traveler out of their way. To arrive at the penalty expressed in Equation (5), subtract the travel time saved ΔT_1 for a particular trip (Eq. 4) from the maximum travel time saved ΔT_{max} (Eq. 3):

(5) *Travel-time penalty*

$$P_1 = \Delta T_{max} - \Delta T_1 \geq 0$$
$$P_1 = (C_{between} - M) - (C_{direct} - \{C_1 + M\})$$
$$P_1 = C_{between} - C_{direct} + C_1$$

In the above equations, M equals all of the components of a maglev-assisted trip other than car travel to and from maglev stations: M = W + T + TP + LH. An analogous equation to Equation (5) could be stated for $C_1 = 0$ and $C_2 > 0$.

To a good approximation, the travel-time penalty associated with a point depends exclusively

Table 1: Data considered in establishing travel time $C_{between}$ by car between maglev stations.[a]

Traffic level	Start time	$C_{between}$	Bing Maps Route API		Bing Maps web page		Google Maps web page	
			NB	SB	NB	SB	NB	SB
Travel between Baltimore's Camden Yards and Washington's Mount Vernon Square								
Rush hour	5 p.m. Monday	70	78	68	76	69	50–75	50–70
Light traffic	8 a.m. Sunday	50	49	48	49	49	40–60	35–55
Travel between BWI Airport and Washington's Mount Vernon Square								
Rush hour	5 p.m. Monday	63	70	59	68	59	40–70	40–65
Light traffic	8 a.m. Sunday	43	43	43	43	43	35–55	35–50

[a] All values are in minutes. "NB" and "SB" stand for "northbound" and "southbound" trips between the two cities. The Bing Maps Route API is described in Appendix 1 of the present chapter. The interactive Bing Maps and Google Maps web pages are https://www.bing.com/maps and https:///www.google.com/maps.

on this point's relation to the nearest maglev station. This penalty is largely independent of where the other terminus of the trip is in relation to the other maglev station. The reason is that most trips between Baltimore and Washington go through the same segment of Interstate 95, a segment that lies between the two cities.

Equation (6) gives an expression for travel time saved for a particular trip origin-destination pair as a function of the travel-time penalties at both the Baltimore and Washington ends of the trip. The present chapter evaluates Equation (6) using a maglev station at Camden Yards and evaluates the equation again using a maglev station at BWI airport. The value most favorable to the maglev is selected as the travel time saved ΔT for this origin-destination pair.

(6) *Travel time saved*

$$\Delta T_{Camden} = \Delta T_{max,Camden} - (P_{Camden} + P_{DC})$$
$$\Delta T_{BWI} = \Delta T_{max,BWI} - (P_{BWI} + P_{DC})$$
$$\Delta T = \max(\Delta T_{Camden} , \Delta T_{BWI})$$

On page 58, Table 3 proves that Equation (6) is consistent with the definition of travel time saved, i.e., Equation (2).

Figures 6, 7, and 8 are the travel-time penalty maps for three scenarios: rush hour if using the Camden Yards maglev station, rush hour if using the BWI maglev station, and light road traffic using the Camden Yards maglev station. No map is generated for light road traffic using BWI because, to a first approximation, no trips ending outside of the airport terminal itself can save the customer 8 to 27 minutes of travel time when road traffic is light.

Table 2. Maximum travel time saved ΔT_{max} and the variables that define its value.[a]

Road traffic and maglev frequency	W[b]	Time to drive to/from Mt. Vernon Sq., $C_{between}$ (quantity M)[c]		Maximum travel time saved, ΔT_{max}[d]	
		Camden Yards	BWI Airport	Camden Yards	BWI Airport
Rush hour and 8 minutes between departures	4	70 (33=4+6+4+15)	63 (30=4+6+4+12)	41	37
Maximum service during light road traffic: 15 minutes between departures	7.5	50 (36.5=7.5+6+4+15)	43 (33.5=7.5+6+4+12)	17.5	13.5
Minimum service during light road traffic: 30 minutes between departures	15	50 (44=15+6+4+15)	43 (41=15+6+4+12)	10	6

[a] All values in the table have units of minutes.
[b] Average wait time for boarding a maglev in minutes.
[c] Based on Table 1, $C_{between}$ is 70 and 50 minutes for rush hour and in light road traffic and $C_{between}$ is 7 minutes less for BWI than for Camden Yards.
[d] ΔT_{max} is 4 minutes less for BWI than Camden Yards because (1) the 7-minute difference in $C_{between}$ and (2) the 3-minute difference in M because the maglev line-haul time is 12 minutes to travel between Washington and BWI vs. 15 minutes to travel between Washington and Camden Yards.

Technical Details of Using the Bing Maps Route API

Equation (5) is evaluated using a Unix shell script that calls the Bing Maps Route application programming interface (API). As of early 2021, the API is called by accessing a URL with the following format: [16]

```
https://dev.virtualearth.
net/REST/V1/Routes/Driving
? o=xml & wp.0=START &
wp.1=END & dateTime=TIME &
optimize=timeWithTraffice &
key=KEY
```

where KEY is the license key that Microsoft has assigned to the user. START and END are latitude-longitude points in the format of north latitude followed by east longitude (e.g., "39.284086,-76.619157"). TIME is a date-time string whose format is mm/dd/yyyy%20HH:MM:SS in which mm is month, dd is day, yyyy is year, HH is hours in the local time zone, MM is minutes, and SS is seconds (e.g., "04/04/2021%2008:00:00"). The Bing Maps API provides its HTTP response in the JSON format. From the JSON output, extract the TravelDurationTraffic field and divide by 60 to convert from seconds to minutes. From a scatter

---

16. Microsoft 2018.

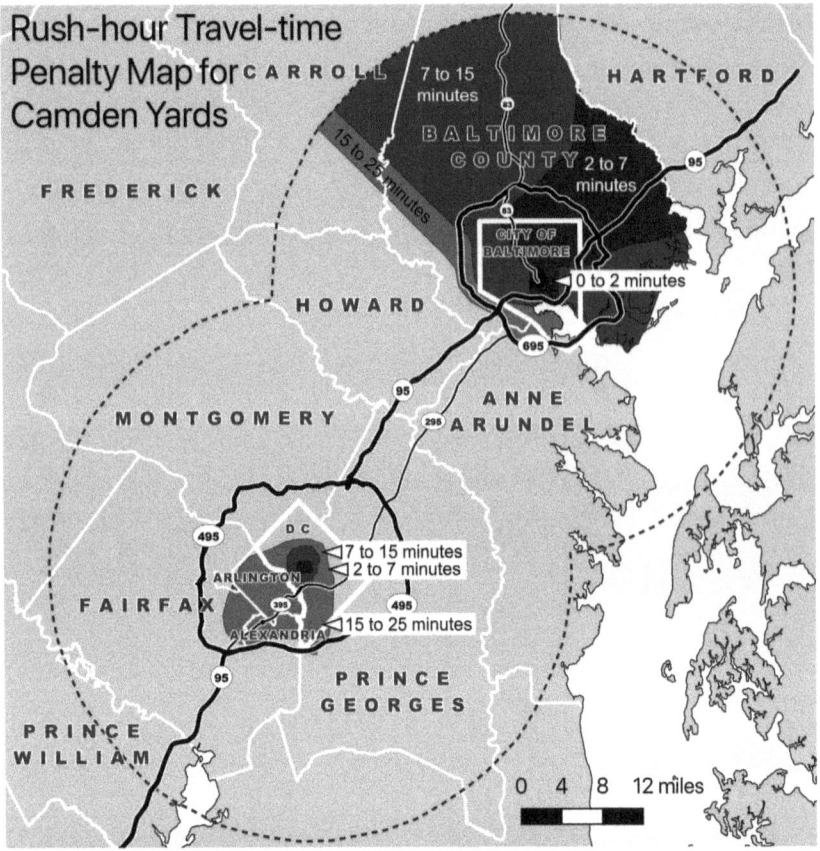

Figure 6. The rush-hour travel-time penalty map when the Baltimore end of the trip uses the maglev station proposed at Camden Yards. The contours of travel-time penalty are defined by Equation (5) and discussed in Appendix 1 of the present chapter. The penalty in minutes represents how far out of the way the maglev stations are for people traveling between Baltimore and Washington.

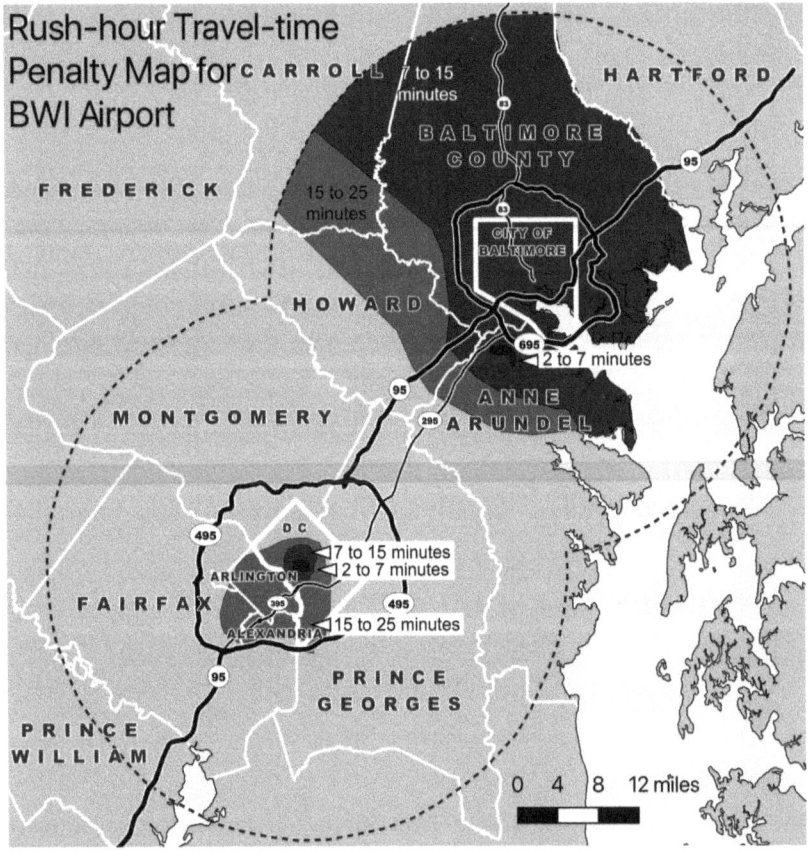

Figure 7. A different realization of the rush-hour travel-time penalty map. The penalty contours in this map differs from those in Figures 6 because, to generate this map, the traveler is assumed to use the maglev station proposed at BWI airport instead of the one proposed at Camden Yards.

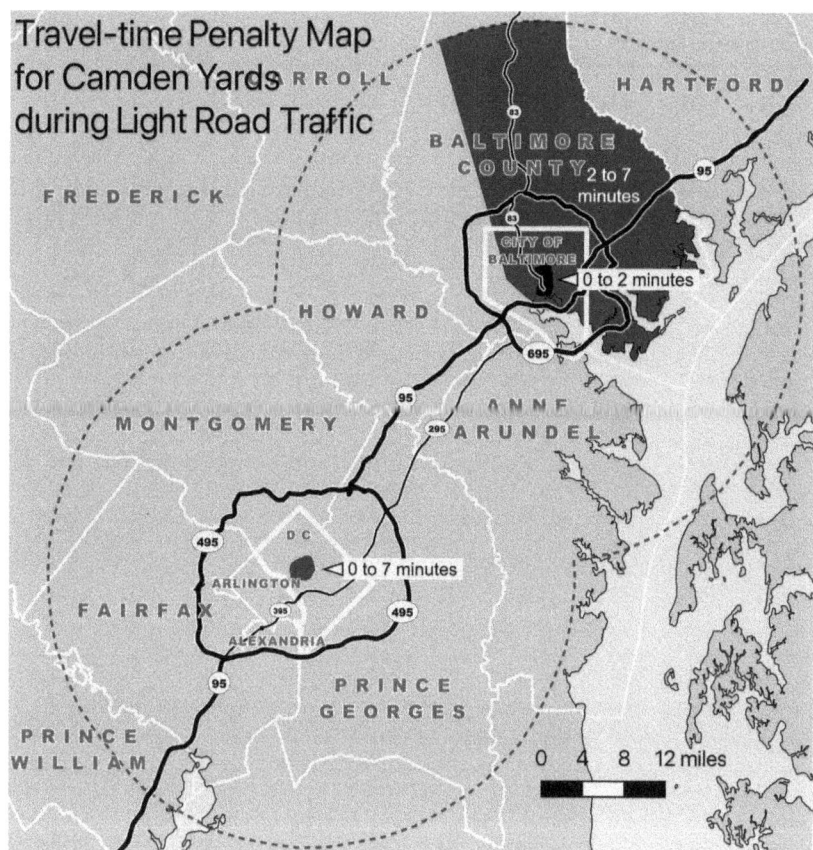

Figure 8. The maglev travel-time penalty map that applies when road traffic is light and the traveler uses the maglev station proposed at Baltimore's Camden Yards. The penalty contours in this map are generated in the same way as those in Figures 6 except that this map is generated for times when road traffic is light.

of trip origin and destination points, identify contours of travel-time penalty and display these contours in QGIS.[17]

The Bing Maps documentation does not describe Microsoft's method for estimating the average duration of car travel at various times of day and days of the week. It appears that, for trips at least one week in the future, long-term average driving conditions dominate rather than observed, recent traffic conditions.

Incorporating US Census Data into Travel-Time Penalty Maps

The present chapter adjusts the travel-time penalty maps at the Baltimore end of the trip using two ideas stated in the maglev's draft environmental impact statement (DEIS).

First, the DEIS states that most maglev customers would start and end their trips within 25 miles of a maglev station to a first approximation. For this reason, the present chapter excludes from the maglev ridership area the portions of Baltimore County and Carroll County that are further than 25 miles from the downtown Baltimore maglev station.[18]

Second, the DEIS states that the 25-mile radius should be expanded or contracted based on where there is an appreciable number of people who currently travel between Baltimore and Washington. Based on this idea, the present chapter excludes Hartford County after consulting US Census data. The 2011–2015 American Consumer Survey (ACS) of the US Census reported that an

---

17. Microsoft 2018; https://www.qgis.org/en/site/.

18. 25-mile radius: Appendix D2, pg. C-106.

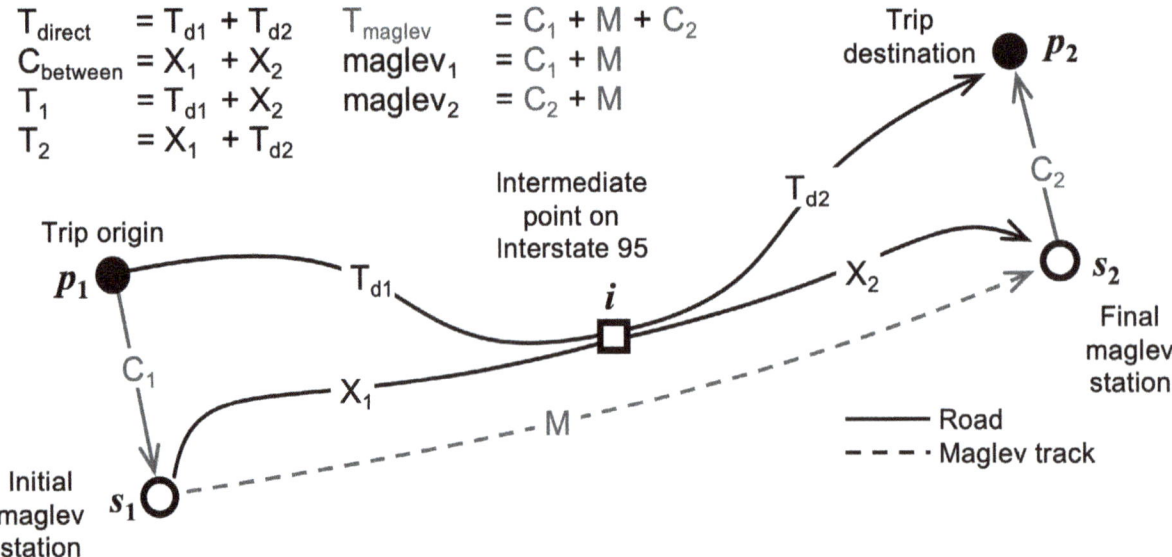

Table 3. A proof that Equation (6) for travel time saved is consistent with the definition of travel time saved, i.e., Equation (2).

Statement	Justification
1. $\Delta T = \Delta T_{max} - (P_1 + P_2)$	The equation to be validated (Eq. 6)
2. $\Delta T = \Delta T_{max} - (\Delta T_{max} - \Delta T_1) - (\Delta T_{max} - \Delta T_2)$	Substitute the definition of travel-time penalty at both ends of the maglev trip, P_1 and P_2
3. $\Delta T = \Delta T_1 + \Delta T_2 - \Delta T_{max}$	Cancel $\Delta T_{max} - \Delta T_{max}$
4. $\Delta T = (T_1 - maglev_1) + (T_2 - maglev_2) - \Delta T_{max}$	Substitute the definition of travel time saved for trips between point p_1 and maglev station s_2 or between point p_2 and maglev station s_1
5. $\Delta T = (T_{d1} + X_2) - (C_1 + M) + (T_{d2} + X_1) - (C_2 + M) - \Delta T_{max}$	Substitute the duration of car travel (T_1) and maglev-assisted trip ($maglev_1$) from point p_1 to maglev station s_2 or of car travel (T_2) and maglev-assisted trip ($maglev_2$) from point p_2 to station s_1
6. $\Delta T = (T_{d1} + T_{d2}) - (C_1 + C_2 + M) + (X_1 + X_2 - M) - \Delta T_{max}$	Reorder terms
7. $\Delta T = T_{direct} - T_{maglev} + (C_{between} - M) - \Delta T_{max}$	Substitute definitions of T_{direct}, T_{maglev} (Eq. 1), and $C_{between}$
8. $\Delta T = T_{direct} - T_{maglev}$	Cancel out $\Delta T_{max} = C_{between} - M$ to obtain the definition of travel time saved, Equation (2)

insignificant number of commuters travel between Hartford County and Washington.[19]

Appendix 2

Using the Travel-Time Penalty Maps to Estimate the Maglev Ridership Area

The goal is to identify the counties and cities within the ridership area of the proposed Baltimore-Washington maglev. Ridership area refers to the area surrounding each station that contains the trip-origin and trip-destination points of most maglev customers. The present chapter identifies the ridership area based on how much travel time could be saved by riding the maglev.

Specifically, this appendix describes a method for identifying the maglev ridership areas as where the maglev would save a traveler approximately 8 to 27 minutes relative to the time that the traveler would otherwise spend driving directly to his or her destination. Examining travel time saved is a valid way to model if a traveler would choose a transportation mode that is more expensive than other options. Serving as a guide for the present chapter, the maglev's draft environmental impact statement (DEIS) simulates a customer's decision to ride the maglev based on whether the customer would find the travel time saved to be worth the cost of the ticket. The present chapter uses the DEIS's 8-to-27-minute range for travel time saved because the author does not know of a compelling reason to use a different range for travel time saved.[20]

The DEIS describes the area where it looked for potential maglev customers but does not mention where its analysis found most maglev riders would start and end their trips. The DEIS ridership model was run over an area approximately defined by a 25-miles radius from each maglev station.[21]

Under different levels of road congestion, this appendix uses travel-time penalty maps (Figures 6 to 8) to generate maps of the maglev ridership area (Figures 2 to 4, page 45).

Travel time saved with a maglev-assisted trip may be represented as the maximum travel time saved minus the sum of travel-time penalties at the Baltimore and Washington ends of the trip, as calculated by Equation (6). For rush hour, these results are shown in two travel-time penalty maps, Figures 6 and 7. There is a separate map for when the traveler uses the maglev station proposed for Baltimore's Camden Yards (Figure 6) and another when the traveler uses the maglev station proposed for BWI airport (Figure 7).

If one end of the trip is near the maglev station in one city then the other end may extend further from the maglev station in the other city with the stipulated travel-time savings still being achieved. In other words, there are options for how to allocate the available "time penalty" between the Baltimore and Washington ends of the maglev-assisted trip. Two such options are explored below.

---

19. ACS data download: US Census Bureau 2015; ACS documentation: US Census Bureau 2017; Example of use of ACS commuter data: Roy 2017.

20. Time vs. cost: Willumsen 2014, pg. 77; trip characteristics vs. implied value of time: Appendix D2, pg. D-107.

21. 8–27 minutes: Appendix D4, pg. C-6; 25-mile radius: Appendix D2, pg. C-106; urban core: Appendix D4, pg. C-6.

The Maglev Ridership Area During Rush Hour: Maximizing Washington-Area Locations Reached

One option is to use most of the available time penalty at the Washington end of the trip to expand the maglev ridership area there. As will be shown, this option permits up to 25 and 8 minutes of travel-time penalty at the Washington and Baltimore ends of the trip, respectively. In this analysis, the customer may use either the maglev station proposed at Baltimore's Camden Yards or BWI airport, just south of the city.

First, consider trips that use the maglev station proposed for Camden Yards. To check that the penalty budget has been correctly applied, use Equation (6), subtracting the penalties for the two cities (25 and 8 minutes) from the maximum travel time saved (41 minutes) in Table 2 to arrive at a minimum savings (8 minutes) that is close to the bottom of the target range of 8 to 27 minutes of savings. The resulting equation is 8 = 41 - (25 + 8), which validates that these two travel-time penalties (25 and 8 minutes) are appropriate for the target amount of travel time saved. Next, look up the geographic extent of these two travel-time penalties or the closest penalties drawn in Figure 6 on page 56. The closest penalty contours drawn are the 25-minute and 7-minute contours at the Washington and Baltimore ends of the trip, respectively.

The just-described process can be repeated for BWI. Specifically, consider trips using the maglev station proposed next to the main terminal of BWI airport. In this case, at least 3 minutes of travel time can be saved using the 25-minute and 7-minute penalty contours in Figure 7. Again, these values can be checked with Equation (6): 5 = 37 - (25 + 7).

Combining the areas identified in the two preceding paragraphs, one finds that the maglev ridership area during rush hour includes about half of the District of Columbia; most of Arlington County and the cities of Alexandria and Baltimore; and less than half of Baltimore County suburbs. Small areas in westernmost Fairfax County and adjacent to BWI airport can also be reached. Figure 2 (page 45) shows this realization of the combined Camden Yards and BWI rush-hour maglev ridership area.

The BWI maglev station contributes a small portion of the maglev ridership area because the neighborhoods near BWI are well served by Interstates 95 and 295, routes that go directly to Washington. The fence around the BWI runways blocks quick access to the BWI maglev station except from the west.

The Maglev Ridership Area During Rush Hour: Maximizing Baltimore-Area Locations Reached

Instead of maximizing locations reached at the Washington end of the trip, one could maximize locations at the Baltimore end. To do so, pick different contours in the rush-hour penalty maps (Figures 6 and 7). One needs to preserve a large enough area in Washington that at least some people would find the trip worth making. For example, using a 7-minute penalty in Washington results in an essentially useless set of trips. In contrast, a 15-minute penalty contour in Washington gives a small but reasonable set of destinations. Using a 15-minute penalty in Washington allows only a 15-minute penalty in Baltimore if the 8-to-27-minute travel-time savings target is maintained.

Equation (6) can be used to verify that using a 15-minute travel-time penalty at both ends of the trip is consistent with overall savings of about 8 to 27 minutes of travel time. Using the Camden Yards maglev station, Equation (6) gives 11 = 41 - (15 + 15), and using the BWI maglev station, Equation (6) gives 7 = 37 - (15 + 15).

On page 56, Figure 7 shows that the BWI station, not the Camden Yards station, would service a small part of eastern Carroll County and northern Anne Arundel County. If the BWI maglev station were taken out of service or became painfully congested because of maglev customers, then the maglev ridership area would shrink and it would no longer contain these portions of Carroll and Anne Arundel Counties. This portion of Carroll County is sparsely populated and this part of northern Anne Arundel County includes Glen Burnie and Pasadena.

While it seems impressive at first glance to have such a large ridership area at the Baltimore end of the trip in Figure 7, few people would find this set of trips useful because so much of the District of Columbia cannot be reached. The excluded portions of the District of Columbia include the Capitol building, Capitol Hill, most of the District's residential areas, and the federal offices just south of the National Mall.

The Maglev Ridership Area when Road Traffic is Light

Compared to rush hour, the biggest difference in light road traffic is that car travel is much faster. Another important difference is that the frequency of maglev trains is less, with the average wait time between trains increasing from 5 to either 7.5 or 15 minutes.

First, consider the maximum frequency of maglev service outside of rush hour, which would be maglev departures every 15 minutes in each direction. In this case, Table 2 gives the theoretical maximum travel-time savings for the Camden Yards and BWI maglev stations as 17.7 and 13.4 minutes, respectively. After subtracting 7 minutes for a Washington penalty and 8 minutes of DEIS-specified travel-time savings, the result is a very small or non-existent amount of travel-time penalty available at the Baltimore end of the trip. Specifically, 2.7 minutes or zero minutes for travel that uses the Camden Yards or BWI maglev stations, respectively.

This result can be checked using Equation (6): $8 = 17.7 - (7 + 2.7)$ and $8 = 13.4 - (7 + x)$ where x cannot be positive. For Camden Yards, the closest-matching penalty contour in Figure 6 is selected (page 56), which is the 2-minute contour. The result is the light-traffic maglev ridership area shown in Figure 4 (page 47). The BWI maglev station would have no utility outside of rush hour beyond providing access to the BWI airport terminal itself. For this reason, no penalty map needs to be drawn for the BWI maglev station when road traffic is light.

Summarizing the above results, the maglev ridership area when road traffic is light is at best a small portion of downtown Washington and downtown Baltimore, as shown in Figure 4.

There are three aspects worth noting about the maglev's limited utility when road traffic is light. For one thing, the travel time saved in light road traffic is near the bottom of the target range, which is the DEIS-specified 8-to-27-minute range. Using the Camden Yards maglev station, the possible travel-time saving is 10.5 to 17.5 minutes ($\Delta T \in [10.5, 17.5] = 17.5 - [2+5,0]$). This is a disappointing amount of travel time saved, suggesting that only wealthier maglev customers would find the travel-time savings sufficient to justify the $40–to–$80 maglev ticket price when road traffic is light.

Second, the maglev ridership area in light traffic contains only a small portion of downtown Washington and downtown Baltimore. The area excludes most of the District of Columbia's residential areas and the federal offices south of the National Mall. It excludes part of downtown Baltimore and all Baltimore County suburbs.

Third, this ridership area outside of rush hour exists only at the maximum maglev service level outside rush hour, i.e., departures every 15 minutes in both directions. At the minimum service level (departures every 30 minutes in each direction), the maglev ridership area is essentially non-existent.

Appendix 3

The Washington Subway's Limited Impact on the Maglev Ridership Area

This appendix shows that, to a first approximation, the maglev ridership area may be identified without reference to the Washington subway, locally known as the "Metro." In other words, this appendix establishes that it is a valid simplification for Appendix 1 of the present chapter to model maglev customers as traveling by car, not subway, to and from the maglev station proposed at Mount Vernon Square in downtown Washington.

Subway-Trip Duration

The total duration D_{subway} (minutes) of a subway trip is the sum of the following terms:

(7) *Travel time by subway to the maglev station in downtown Washington*

$$D_{subway} = F_s + T_s + W_s + LH_s$$

where F_s is the time to walk or drive from the trip origin to the subway-station entrance. T_s is the sum of the time to walk through the initial subway station to reach the platform and to walk from the final subway station's platform to the entrance of the maglev station. W_s is the average time spent waiting to board the initial subway train. LH_s is the line-haul time for the subway trip, i.e., the time between boarding the subway train at the initial subway station and disembarking the subway train at the final subway station.

Equation (7) calculates the duration of a subway trip in one direction. However, the trip duration would be essentially the same if the subway trip were made in the reverse direction, i.e., starting at the Mount Vernon Square maglev-station entrance and riding the subway toward the trip destination.

In calculating the values in Table 4 on page 65, F_s is set to zero. In other words, Table 4 shows the travel time if one starts the trip exactly at the entrance of a subway station. Once a subway station of interest has been identified, one can increase F_s until the total travel time reaches the maximum value that still allows the traveler to save the DEIS-specified 8 to 27 minutes of travel time on the maglev-assisted trip.

The time to walk through the initial and final subway stations is about 3 minutes for each station, so the total for both ends of the subway trip would be about 6 minutes for T_s. This estimate is based on Appendix 1 of the present chapter that found a walking time of 3 minutes in a maglev station. Maglev stations are roughly similar in size to Washington subway stations. Equation (7) makes the simplification that the Mount Vernon Square maglev-station entrance is immediately adjacent to the Mount Vernon Square subway-station entrance.

The Washington Metropolitan Area Transit Authority (WMATA), which runs the Washington subway system, stated as of March 2021 that the time between subway trains during weekday rush hour is 6 minutes on the Red Line and 12 minutes on the other lines. Taking half the time between train departures as the average wait time, W_s will have a value of 3 to 6 minutes depending on which subway line is used.[22]

Subway line-haul times may be estimated using WMATA's trip-planner web page. The line-haul time may include intermediate station stops and a transfer from one subway line to another. In Table 4, transfer time is assumed to be half of the time between trains on the line the passenger is transferring to.[23]

Three Criteria for Identifying Which Subway Stations Would Expand the Maglev Ridership Area

There are three criteria that a subway station must meet if it is to expand the maglev ridership area beyond the maglev ridership area calculated in Appendix 1 assuming car travel takes people to and from maglev stations.

The first criterion is that riding the subway to the maglev station must take less time than driving there. Stations that meet the first criterion can be identified in Table 4 as stations where the minutes in the Subway column are lower than the range of minutes in the "Car (Bing)" column. Stations that meet this criterion have their entry in the Subway column in square brackets.

The second criterion is that the subway station must be outside of the rush-hour maglev ridership area calculated in Appendix 1. The square brackets around entries in the Penalty column of Table 4 identify stations that meet this criterion. Specifically, the square brackets are used for travel-time penalties of up to 10 minutes greater than the 25-minute limit in Washington rush hour discussed in Appendix 2.

The third criterion is that the maglev-assisted trip using the subway must save the DEIS-specified 8 to 27 minutes of travel time. Table 4 can be used to identify which stations satisfy the third criterion during rush hour. One calculates how many minutes faster taking the subway to the maglev station is compared to driving to the maglev station (subtract the Subway column from the "Car (Bing)" column). One determines whether this difference is at least as great as the number of minute that the Penalty column is greater than the 25-minute limit discussed in Appendix 2.

Two stations meet all three criteria: PG Plaza on the Green Line and Franconia-Springfield on the Yellow Line. Their names are in square brackets in the leftmost column of Table 4 to identify that these stations meet the criteria. Examples are given below.

Applying the Three Criteria to Specific Subway Stations

This section works out two rush-hour examples and explains why non-rush-hour examples are not needed. The examples are for the Washington subway contribution to the maglev ridership area when that area is optimized to increase coverage at either the Washington or Baltimore end of the trip. These two optimization options are shown in Figures 2 and 3, as previously discussed in the present chapter.

The first example applies when optimizing the maglev ridership in Washington during rush hour, as shown in Figure 2. In this case, the only subway stations that satisfy all three criteria specified in the previous section are PG Plaza and Franconia-Springfield. The rush-hour maglev ridership area would expand only slightly by including a

---

22. Time between subway train departures are found on the timetable: https://www.wmata.com/schedules/timetables/.

23. Time between stations: https://www.wmata.com/schedules/trip-planner/.

small area around the PG Plaza and Franconia-Springfield subway stations.

The detailed application of the three criteria is as follows. During rush hour, riding the subway from PG Plaza to the Mount Vernon Square maglev station would save 15 minutes relative to making that trip by car. The same is true for the Franconia-Springfield subway station except that only 10 minutes would be saved instead of 15. Driving from the PG Plaza or Franconia-Springfield subway stations to the maglev station would add 9 or 4 minutes beyond the 25-minute penalty contour. Combining this information, the net travel-time savings using the subway is only 6 minutes (6 = 15 - 9; 6 = 10 - 4). During rush hour, it isn't possible to start a trip very far from the PG Plaza or Franconia-Springfield subway station and reach the subway station in 6 minutes. It appears that only about a 1-mile radius from each of the two subway stations can be reached in this way. This 1-mile radius would result in a small expansion beyond the maglev ridership area that assumes car travel to and from maglev stations.

The second example applies when optimizing the maglev ridership in Baltimore during rush hour, as shown in Figure 3. In this case, a 15-minute travel-time penalty in Washington was used to maximize the maglev ridership area in Baltimore in Appendix 2 of the present chapter. In this scenario, the Pentagon and Columbia Heights subway stations barely satisfy all three criteria, leaving no time to walk or drive to them. For this reason, it appears these stations would add such a small area to the maglev ridership map that this area would not be easily visible.

So far, the examples have all been during rush hour. Outside of rush hour, car travel is faster and subway trains less frequent. For these reasons, the subway does not expand the maglev ridership area at all outside of rush hour.

To summarize, the preliminary analysis presented here suggests that subway travel to or from the Mount Vernon Square maglev station would create only a small expansion of the maglev ridership area, and it would do so only during rush hour. To a first approximation, it is safe to calculate the maglev ridership area under the simplifying assumption that people travel by car to and from a maglev station, not by subway.

Table 4. Trip duration in minutes during rush hour to travel to the proposed downtown Washington maglev station at Mount Vernon Square. Travel to the maglev station is either by subway or by car.[a]

Initial subway station or start of the drive to the maglev station	Subway line-haul time			Travel time to maglev station			Penalty[f]
	1st leg	Transfer[b]	2nd leg	Subway[c]	Car[d] (Google)	Car[e] (Bing)	
Green Line							
Greenbelt	26	--	--	[35]	22–45	39–48	79
[PG Plaza]	18	--	--	[27]	18–45	41–42	[34]
West Hyattsville	15	--	--	[24]	16–40	33–35	22
Fort Totten	12	--	--	[21]	12–30	27–33	16
Columbia Heights	5	--	--	14	7–16	15–22	19
Shaw	1	--	--	10	3–8	6.5–7.5	8
Gallery Place	2	--	--	11	2–6	4.1–4.9	4
L'Enfant Plaza	6	--	--	15	5–14	12–13	19
Naylor	21	--	--	30	14–30	22–28	[31]
Yellow Line							
Pentagon	12	--	--	21	12–22	20–29	19
Alexandria	21	--	--	30	16–30	34–40	27
Huntington	27	--	--	36	20–35	33–38	[27]
[Franconia]	29	--	--	[38]	24–45	46–51	[29]
Red Line transferring to Green or Yellow Line at Gallery Place							
Woodley Park	8	3	2	19	8–22	17–19	17
Farragut North	3	3	2	14	6–16	12–13	3
Metro Center	2	3	2	13	4–10	7.3–7.7	7
Union Station	5	3	2	16	7–18	11–18	16
Orange Line transferring to Green or Yellow Line at L'Enfant Plaza							
Vienna	37	3	6	58	28–40	39–40	[35]
Federal Center	7	3	6	28	6–16	13–14	19

[a] The square brackets are explained in Appendix 3 of the present chapter.

[b] The average time is 3 minutes to transfer to a train on the Green or Yellow Lines at Gallery Place or L'Enfant Plaza. Assuming each of these two lines has a train departing every 12 minute, that means a departure every 6 minutes and an average wait time of half that (3 minutes).

[c] As defined in Equation (7), the total time to travel by subway from the entrance of the subway station listed in Table 4's leftmost column to the entrance of a maglev station at Mount Vernon Square. To calculate total subway travel time, add 3+3 for T_s+W_s to the line-haul time if the trips starts on the Red Line, or add 3+6 if the trip starts on any other subway line.

[d] The range of car-travel trip durations provided by interactive Google Maps web page, https://www.google.com/maps.

[e] Car-travel trip duration provided by the Bing Maps Route API. A range is stated in the table because the API calculates a different trip duration depending on the direction of travel.

[f] Maglev travel-time penalty at the entrance of the subway station stated in the leftmost column of Table 4, as calculated by Equation (5) of Appendix 1 of the present chapter.

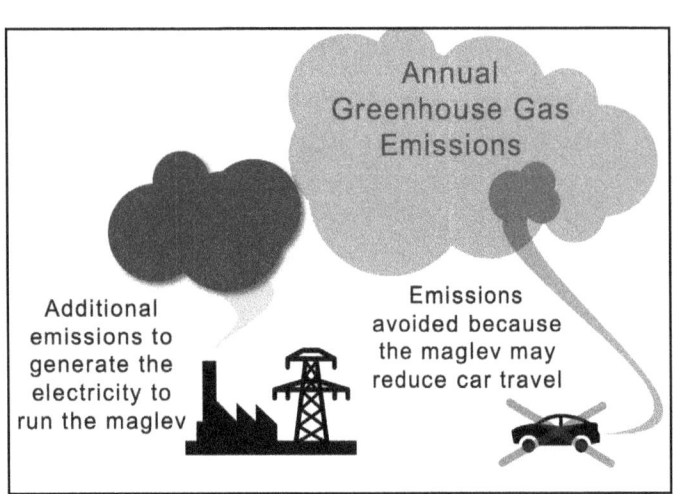

GREENHOUSE GAS • PAGE 69

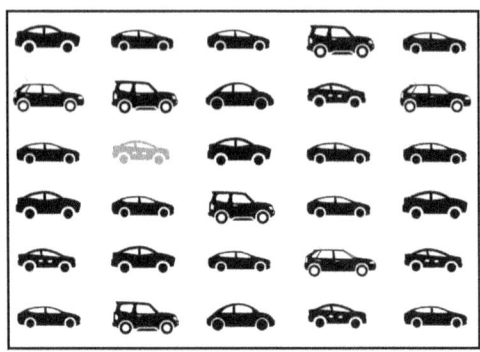

ROAD CONGESTION
PAGE 83

The train schedule stated in the draft environmental impact statement determines the greenhouse gas emissions from operating the proposed maglev. These increased emissions would be partially offset by the forecasted reduction in greenhouse gas emissions because some car drivers would switch to riding the maglev. If the advertised train schedule produces less than the forecasted number of car drivers switching to the maglev, then the maglev's net climate impact would be worse than forecasted.

Most maglev riders are expected to be diverted car travelers, i.e., people who would travel between Baltimore and Washington by car if the maglev were not built. If maglev ridership was less than the official forecast, then the maglev would do even less to reduce regional road congestion than is predicted in the draft environmental impact statement.

PART TWO

Maglev Impacts that Vary with the Ridership Forecast

```
an extraordinarily costly but also an abnormally energy-
wasting project
```

> —Hidekazu and Nobuo (2017), in "End Game for Japan's Construction State-The Linear (Maglev) Shinkansen and Abenomics." The Linear Shinkansen is a term used in Japan for the technology that is proposed for the Baltimore-Washington maglev.

```
Maglev has all the defects of conventional high-speed
rail with the added bonuses of higher costs and greater
energy requirements.
```

> —Randal O'Toole, Cato Institute *At Liberty* blog (2013 Nov 06)

4 • GREENHOUSE GAS

Operating the proposed Baltimore-Washington maglev would increase greenhouse gas emissions, the Federal Railroad Administration finds

According to the project's draft environmental impact statement, operating the maglev would increase annual carbon dioxide emissions by more than a hundred million kilograms, contradicting the claims of maglev promoters

---

The plot has thickened regarding the climate-change impact of the "maglev"—the superconducting magnetic-levitation rail line that has been proposed to connect Baltimore and Washington, DC. In December 2020, the present author estimated that constructing the maglev would release hundreds of millions of kilograms of carbon dioxide. This earlier analysis was published in the Issues Forum of the Prince George's County group of the Sierra Club.

Since then, a regulatory agency has published an analysis covering the other side of the question: operation rather than construction. How much would operating the maglev increase annual carbon-dioxide emissions? The regulatory agency's findings, however, are being ignored by some companies and news organizations.

This unfortunate situation will be discussed and clarified in this chapter.

In January 2021, this regulatory agency—the Federal Railroad Administration—published the draft environmental impact statement for the proposed maglev. It indicated that operating a maglev between Baltimore and Washington would increase annual carbon-dioxide emissions by more than a hundred million kilograms because of the large amount of electricity that the maglev would consume. This part of the impact statement directly contradicted claims broadcast for years by the company that wants to build the maglev, Baltimore Washington Rapid Rail (BWRR).[1]

Ignoring the draft environmental impact statement, BWRR and its parent company, The Northeast Maglev, continue to repeat their claims

1. Emission increase due to maglev operation in the DEIS (FRA 2021), Appendix D4, as described in the present chapter. Statements by BWRR and TNEM at https://bwrapidrail.com, https://northeastmaglev.com.

that the maglev would reduce greenhouse gas emissions. Several newspapers mentioned later in this article have echoed the companies' claims. Such reporting serves to hide from public view the greenhouse-gas findings of the Federal Railroad Administration.

The Federal Railroad Administration bears some responsibility for this situation because of several editorial choices made in the draft environmental impact statement that the agency managed, reviewed, approved, and published. Specifically, the statement buries greenhouse-gas findings in an appendix and makes no mention of them in the document's Executive Summary.

What We Knew Before 2021

To understand the draft environmental impact statement published in 2021, it helps to review prior years' statements about the proposed maglev's greenhouse gas impact.

In 2015, Wayne L. Rogers, the chairman of Baltimore Washington Rapid Rail, testified before the Maryland Public Service Commission that the maglev would reduce greenhouse gas emissions by 2 million short tons (1,814 million kilograms). Rogers said this figure came from a report authored by Louis Berger, a consulting company.[2]

A summary of the Louis Berger report was also submitted as evidence in this 2015 case. The 2 million short tons quoted by Rogers turned out to be an estimate for the entire lifetime of the project, not the per-year emission savings. The Louis Berger report summary states an estimate of carbon dioxide emissions from operating the maglev over the project's lifetime with no mention of the emissions that would result from constructing the maglev in the first place. Louis Berger started with an estimate of the carbon dioxide (CO_2) emissions to generate the electricity to run the Baltimore-Washington maglev. From this value, the company subtracted its estimate of the CO_2 emissions that would be avoided because of reduced car travel. The reasoning is some travelers would switch from driving to riding the maglev.[3]

On an annual basis, the Louis Berger estimate of CO_2 savings is rather small. One can convert project-lifetime emissions to annual emissions by dividing by 60 years, a value found in the literature. The result is savings of only 33 million kilograms of CO_2 per year.

In comparison, the Maryland Department of Energy estimates that a more significant reduction in annual CO_2 emissions could be achieved, at a much lower cost, by expanding telework opportunities in Maryland: an annual emissions reduction of 300 to 790 million kilograms. This impact would be about ten times greater than the above-mentioned Louis Berger estimate. An even more significant reduction could be achieved in Maryland, again at low cost, by increasing the fuel-economy standard for gasoline-powered cars: a 3,680-million-kilogram reduction per year. This impact would be about a hundred times greater than the impact from the proposed maglev.[4]

Even if the Louis Berger CO_2 emission estimate were accurate, it would still take the maglev about a decade or two to cancel out

---

2. Rogers 2015, pg. 19; 1 short ton is about 907.2 kilograms, and 1 metric ton is exactly 1,000 kilograms.

3. 2.185 million short tons (1,982 million kilograms): Louis Berger 2015, pg. 7.

4. The Louis Berger report summary did not state the company's estimate for the maglev's lifetime. Kato and Shibahara (2005) used 60 years for the useful life of the maglev track. Maryland DOE (2021) states impact of expanded telework (Table 3.2-8, pg. 103) and car emission standards (Table 3.2-5, pg. 91).

the CO_2 emissions from its construction. In December 2020, the present author published an initial estimate that constructing the maglev track and tunnel between Baltimore and Washington would release hundreds of millions of kilograms of carbon dioxide. This estimate is found, in a slightly revised form, in Appendix 2 of the present chapter. It appears that no other individual or organization has published an estimate for the amount of CO_2 that would be emitted to construct a maglev between Baltimore and Washington.

New In 2021

Contradicting the 2015 Louis Berger report, the draft environmental impact statement published in January 2021 asserted that maglev operation would significantly increase greenhouse gas emissions. Specifics are provided in the next section of the present chapter.

On February 9, 2021, the editorial board of the *Baltimore Sun* published an op-ed that asserted the maglev would reduce greenhouse gas emissions. The paper presented no evidence to support this assertion.

On April 2, 2021, the *Washington Post* published an article claiming that the maglev would "help cut greenhouse gas emissions" because it would take "about 16 million car trips off the road annually." The Post's argument is specious: superficially plausible but flawed.

Contrary to what the *Washington Post* published, the amount of car travel the maglev replaces does not determine whether maglev operation causes a net increase or decrease in greenhouse gas emissions. What determines the sign and magnitude of net emissions is whether generating the electricity to run the maglev would emit more carbon dioxide than would be avoided through the maglev-related reduction in car travel. It also matters how much carbon dioxide would be emitted to construct the maglev track and related facilities. The following two sections examine in greater detail the greenhouse gas impact of maglev operation and construction.[5]

Carbon Dioxide from Operating the Maglev

The bottom line is the draft environmental impact statement (DEIS) indicated that operating the maglev would emit 286 to 336 million kilograms more carbon dioxide each year than if the maglev were not operated. This information is found in Appendix D4 of the DEIS. The mathematical details are explained in Appendix 1 of the present chapter and are shown schematically in Figure 1 on the next page.

The case for building the maglev is weakened because the DEIS determined that greenhouse gas emissions would increase due to maglev operation. The DEIS-identified increase certainly paints the maglev in a different light than the decrease in emissions suggested by the 2015 Louis Berger report that was discussed earlier in the present chapter.

While Appendix D4 of the DEIS shows that maglev operation would increase greenhouse gas emissions, the DEIS contains two misleading statements on this topic.

First, consider Section 16 of Chapter 4 that claims the "FRA did not quantify the power-plant emissions required for [maglev] train operations and facilities" (pg. 4.16-3). In fact, the Federal

---

5. *Baltimore Sun* on 9 Feb 2021; Luz Lazo in the *Washington Post* on 2 April 2021; As of June 2021, The Northeast Maglev website still claims the maglev would reduce greenhouse gas emissions by 2 million short tons.

Federal Railroad Administration estimate of greenhouse gas impact from operating the proposed Baltimore-Washington maglev
in millions of kilograms of carbon dioxide (CO_2) per year

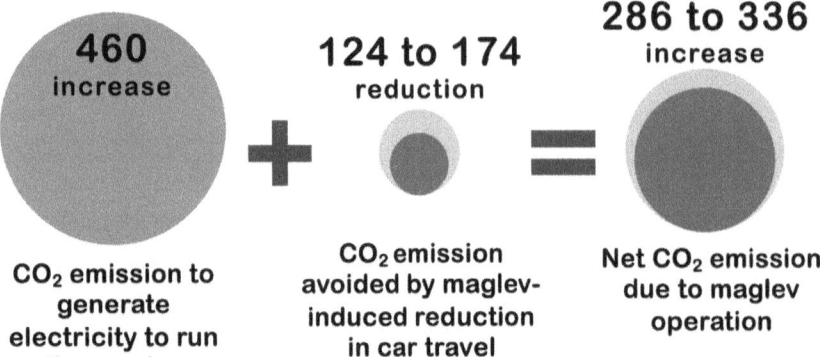

Figure 1. A schematic diagram of how the January 2021 draft environmental impact statement estimates the climate impact of operating the proposed Baltimore-Washington maglev.

From Tables D4-40 and D4-43 of Appendix D4 of the draft environmental impact statement (DEIS) published by the FRA in January 2021 (pages D4-51 and D4-52)

Railroad Administration did provide an estimate for one type of emissions. Specifically, it provided an estimate for CO_2 emissions from power plants providing the maglev its electricity. The agency did so in Table D4-43 of Appendix D4.

Section 16 contains an even more misleading statement: operating the maglev "will not increase greenhouse gas emissions." The two sentences that follow this statement qualify it to the point that it is rendered nearly meaningless. Below is the statement in italics quoted in context:

> The SCMAGLEV system will operate entirely on electricity, with the exception of certain maintenance vehicles. As a result, the SCMAGLEV train *will not increase greenhouse gas emissions*. However, as described in Section 4.19 Energy, the SCMAGLEV system will result in an increase in power consumption in the region. Therefore, an increase in greenhouse gas emissions from power plants would likely occur. (Chapter 4.16, pg. 4.16-11)

It is embarrassing that the Federal Railroad Administration used such tortured logic to insert a misleading statement (the maglev "will not increase greenhouse gas emissions") into the DEIS.

To be clear, it is true that the maglev would decrease CO_2 emissions if one looks only at the forecasted reduction in car travel due to the maglev and one ignores the CO_2 emissions from the electricity generated to run the maglev. While true, this statement is beside the point. The important question is the net effect of maglev operation. This question is addressed in Appendix D4 of the DEIS as discussed in the preceding paragraphs.

The present chapter assumes the official ridership forecast is accurate. If the official ridership forecast was too high, then the net climate impact of maglev operations would be worse. The reasoning is as follows: it would take about the same amount of electricity (and associated CO_2 emissions) to run the maglev trains whether they were full or mostly empty. The number of car vehicle-miles reduced and the predicted CO_2 emission reduction, however, would be smaller if maglev ridership fell short of the official forecast.

Carbon Dioxide from Constructing the Maglev

The DEIS does not quantify the greenhouse gas emissions that would result from manufacturing the material needed to construct the maglev's elevated track, tunnel, and associated facilities. In fact, the DEIS does not even mention that such emissions would occur. A rough estimate is calculated in the present chapter.[6]

A common simplification employed by planners is to estimate the greenhouse gas impact of a construction project based on the emissions to manufacture the concrete and steel required.

Using this method, Appendix 2 of the present chapter derives a lower bound for the CO_2 emissions that would result from maglev construction. Constructing the tunnel and elevated track for the proposed Baltimore-Washington maglev would release 249 to 721 million kilograms of carbon dioxide. This emission range underestimates the total emission from constructing the maglev because it excludes emissions from building the maglev stations, train, control facility, and train-maintenance facility.

If the $17 billion that would be spent to construct the maglev caused carbon dioxide emissions at the average rate for construction projects in the United States, then the total emissions from constructing the maglev would be much higher than the lower bound estimated here. This possibility is discussed in Appendix 2 of the present chapter.

Is the Maglev "Green"?

When evaluating whether or not a project would be environmentally friendly, there is more to consider than just kilograms of carbon dioxide.

In broad terms, the proposed maglev would involve building massive concrete structures, which would decimate green space. It would involve trying to entice people to travel farther and faster at great expense with a significant expenditure of energy. In many ways, such a project would be the opposite of environmentally friendly. Environmental harm, expense, and "induced" travel are each documented in the draft environmental impact statement published in January 2021. Such evidence suggests the maglev isn't green.[7]

A green future is possible for the Baltimore-Washington region. Efforts are being made to realize it. The maglev would do little if anything to contribute to this effort. The specifics follow.

---

6. In Chapter 4.16 starting on pg. 4.16-3, the DEIS explains its CO_2 emission-modeling method. The DEIS emission numbers are found in Appendix D4 pg. D4-51 to D4-53.

7. 15% of maglev trips would be "induced" travel, i.e., travel between Baltimore and Washington that would not occur if the maglev were not built: Appendix D4, Table D4-29, pg. D-45. Construction cost of $15 to $17 billion: Appendix D4, Table D4-8, pg. D-21. Negative impacts would occur to historical sites (Chapter 4.8); scenic resources (Chap. 4.9); recreational facilities (Chap. 4.7); environmental justice (Chap. 4.5); quality-of-life (Chap. 4.4); hazardous waste sites (Chap. 4.15); forests, forest-interior species, and habitats of rare, threatened, and endangered species (Chap. 4.12); wetlands (Chap. 4.11); economic harm during construction (Appendix D4, pg. D-18 to D-30); and lost revenue for Amtrak and MARC commuter trains (Appendix D4, Table D4-47, pg. D-54).

Each community within the region could be strengthened to better meet its residents employment and recreation needs. Such a transformation would reduce the need for long-distance travel across the region, and in turn, would reduce the region's carbon footprint. In addition, expanded options for teleworking could be made available for when interacting with a distant workplace is required. This sort of vision was articulated years ago and has motivated decisions within the Metropolitan Washington Council of Governments. This planning organization has stated the following:

> Locating homes, employment centers, schools, and other activities in closer proximity, and expanding transit, telecommuting, bicycling, and walking options can reduce vehicle miles of travel per capita and improve accessibility throughout the region (MWCOG 2010, pg. 18)

The Council states elsewhere that it intends "expanding transit" to mean expanding transportation options that maximize accessibility and affordability. The proposed Baltimore-Washington maglev would fail to contribute to this goal because the ticket price would be $40 to $80 one way per person.[8]

Conclusion

The Federal Railroad Administration has determined that the proposed Baltimore-Washington superconducting maglev would increase greenhouse gas emissions each year it was operated. This increase is relative to the emissions that would occur if the maglev did not run and people used other transportation options. The draft environmental impact statement (DEIS) quantifies this emission increase on pages D4-51 to D4-53 of Appendix D4. The Federal Railroad Administration published this document in January 2021.

The greenhouse-gas discussion in the DEIS is summarized by the following list:

- Maglev operation would increase net CO_2 emissions by 286 to 336 million kilograms per year relative to the No Build option
- The net CO_2 emissions are the sum of two factors: 460 million kilograms of CO_2 emissions annually from generating the electricity to run the maglev and 124 to 174 million kilograms of CO_2 emissions avoided annually assuming that some car travel would be replaced by maglev travel
- The net CO_2 emissions are not stated explicitly in Appendix D4, but they may be calculated from data found in two tables of Appendix D4
- The DEIS does not estimate the CO_2 emissions from constructing maglev-related infrastructure
- The DEIS Executive Summary makes no mention of the maglev's impact on CO_2 emissions

It is unclear what would motivate the Federal Railroad Administration to de-emphasize its findings regarding the greenhouse-gas emission impact of maglev operation in the DEIS. It is also unclear why the agency did not estimate the greenhouse-gas emission impact that would result from maglev construction.

References

Baltimore Sun, 9 February 2021: What do you call a 311-mph train serving Baltimore? Transformative. editorial board, https://www.baltimoresun.com/opinion/editorial/bs-ed-

---

8. Accessibility quote: MWCOG 2010, pg. 9; ticket price: Appendix D2, pg. D-107, D-108.

0210-maglev-train-baltimore-20210209-lra3n4sgnbe5toyxzbe4ntafca-story.html.

EPA, 2009: *Potential for Reducing Greenhouse Gas Emissions in the Construction Sector.* 50 pp, https://archive.epa.gov/sectors/web/pdf/construction-sector-report.pdf.

Federal Highway Administration, 2018: *Average Vehicle Occupancy Factors for Computing Travel Time Reliability Measures and Total Peak Hour Excessive Delay Metrics.* 8 pp. Available online at https://www.fhwa.dot.gov/tpm/guidance/avo_factors.pdf.

Federal Railroad Administration, January 2021: *Baltimore-Washington Superconducting MAGLEV Project Draft Environmental Impact Statement and Draft Section 4(f) Evaluation.* 3,053 pp. (main text and appendices), https://bwmaglev.info/index.php/project-documents/deis.

Fantilli, A. P., O. Mancinelli, and B. Chiaia, 2019: The carbon footprint of normal and high-strength concrete used in low-rise and high-rise buildings. *Case Studies Construct. Materials*, **11**, https://doi.org/10.1016/j.cscm.2019.e00296.

Gursel, A. P., 2014: *Life-Cycle Assessment of Concrete: Decision-Support Tool and Case Study Application.* Ph.D. diss., UC Berkeley, 541 pp, https://escholarship.org/uc/item/5q24d64s.

Huang, L., G. Krigsvoll, F. Johansen, Y. Liu, and X. Zhang, 2018: Carbon emission of global construction sector. *Renewable and Sustainable Energy Review,* **81**, 1906–1916, https://doi.org/10.1016/j.rser.2017.06.001.

International Energy Agency (IEA), 2019: *The Future of Rail: Opportunities for Energy and the Environment.* 175 pp. Fig. 1.30 on pg. 57. Available online at https://webstore.iea.org/download/direct/2434.

Kato, H., and N. Shibahara, 2005: A life cycle assessment for evaluating environmental impacts of inter-regional high-speed mass transit projects. *J Eastern Asia Soc. Transportation Studies*, **6**, 3211–3224, http://citeseerx.ist.psu.edu/viewdoc/download? doi=10.1.1.556.5875 & rep=rep1 & type=pdf.

Kelley, O. A., 13 December 2020: Would the Proposed Baltimore-Washington Maglev Increase Greenhouse Gas Emission? Issues Forum of the Prince George's County group, Sierra Club, https://www.sierraclub.org/maryland/prince-georges/issues-forum.

Lazo, L., 2 April 2021: A maglev would be a speedy option over protected land. But research and wildlife might suffer. *Washington Post*, https://www.washingtonpost.com/transportation/2021/04/02/maglev-train-dc-baltimore-environmental-impact/.

Liu, et al., 2019: Development of a carbon emissions analysis framework using building information modeling and life cycle assessment for the construction of hospital projects. *Sustainability*, **11**, https://doi/10.3390/su11226274.

Louis Berger, 15 April 2015: Baltimore-Washington SCMAGLEV Project Economic Benefits Overview. Exhibit B of testimony by Albert Racciatti in case #9363, testimony #25d, MD Public Service Commission, 12 pp, https://www.psc.state.md.us/search-results/? q=9363 & x.x=20 & x.y=20 & search=all&search=case.

Maryland Department of Energy, 19 February 2021: *The Greenhouse Gas Reduction Act: 2030 GGRA Plan.* 280 pp, https://mde.maryland.gov/programs/Air/ClimateChange/Pages/Greenhouse-Gas-Emissions-Reduction-Act-(GGRA)-Plan.aspx.

MWCOG, 2010: *Region Forward Vision.* Metropolitan Washington Council of Governments, 72 pp, https://www.mwcog.org/regionforward/.

Priemus, H., B. Flyvbjerg, and B. van Wee (Eds), 2008: *Decision-Making on Mega-Projects: Cost-Benefit Analysis, Planning and Innovation.* Edward Elgar, pg. 7.

Rogers, W. L., 17 April 2015: Direct testimony of Wayne L. Rogers, case #9363, testimony #26b, MD Public Service Commission, 23 pp, https://www.psc.state.md.us/search-results/? q=9363 & x.x=20 & x.y=20 & search=all & search=case.

Sturm, J., et al., 2011: Analysis of cost estimation disclosure in EIS for surface transportation projects. *Transportation*, **38**, 525–544.

World Bank, 2016: CO_2 emissions (kg per PPP $ of GDP). web page, World Development Indicators, dataset EN.ATM.CO2E.PP.GD, https://databank.worldbank.org/reports.

aspx?source=2&type=metadata&series=EN.ATM.CO2E.PP.GD.

Appendix 1

CO_2 from Operating the Maglev

This appendix describes the mathematical details of how the Federal Railroad Administration expressed the greenhouse-gas emission impact of maglev operation. This information is found in Appendix D4 of the draft environmental impact statement (DEIS) published in January 2021.

Increase in CO_2 Emissions Due to Electricity Generation

Table D4-43 of Appendix D4 states that 460 million kilogram per year of CO_2 would be emitted to generate the electricity to run the maglev trains, stations, control facility, and train-maintenance facility. These CO_2 emissions are the product of two variables: the amount of electricity used and the CO_2 emission rate for the generating facility.

Table D4-43 arrives at the CO_2 emissions from electricity generation in the following way: 100,322 megawatt-hours of Washington power that emits 0.1991 metric tons of CO_2 per megawatt-hour plus 1,161,678 megawatt-hours of Maryland power that emits 0.3791 metrics tons of CO_2 per megawatt-hour. These values assume the downtown Baltimore maglev station is located at Camden Yards, but the values would be similar if the station were in Cherry Hill. This chapter refrains from endorsing the DEIS's per-megawatt-hour rates. Kelley (2020) reported that a somewhat higher emission rate would be more appropriate in this situation, which would increase the CO_2 released by the generation of the electricity to run the maglev.[9]

Decrease in CO_2 Emissions Due to Reduced Car Travel

Table D4-40 of Appendix D4 states that \$124,431 to \$348,536 of CO_2 emission savings would accrue annually because of the forecasted number of people switching from driving cars to riding the maglev. The bottom of this range is determined by the Cherry Hill track alignment in 2030, a scenario in which the DEIS values CO_2 at \$1 per metric ton. The top of this range is determined by the Camden Yards track alignment in 2045, at which time the DEIS values CO_2 at \$2 per metric ton.[10]

The present chapter does not comment on whether the DEIS per-ton cost is the true social cost of CO_2 emissions, but merely identifies this rate as the one used in the DEIS.

Using the DEIS conversion rates, the dollar savings explicitly stated in the DEIS imply 124 to 174 million kilograms of CO_2 emissions savings due to the forecasted maglev-related reduction in car travel.

The just-stated DEIS estimate of the maglev's ability to reduce car-related emissions is more than double the car-related emission reduction estimated by Kelley (2020). The difference can be attributed to two factors. The DEIS assumed that the maglev would divert more car travel than Baltimore Washington Rapid Rail (BWRR) thought possible in 2015. Writing before the

---

9. 1 metric ton per megawatt-hour is equal to 1 kilogram per kilowatt-hour; annual emission from electricity generation to run maglev: 460×10^6 kg y^{-1} = 100.3×10^6 kW·h (y^{-1}) · 0.1991 kg (kW·h)$^{-1}$ + $1,162 \times 10^6$ kW·h (y^{-1}) · 0.3791 kg (kW·h)$^{-1}$; data source: Appendix D4, Table D4-43, pg. D-52.

10. \$1 per ton vs. \$2 per ton cost: Appendix D4, Table D4-43, pg. D4-52.

publication of the DEIS, Kelley (2020) used BWRR's 2015 figures. The DEIS also assumed that gas-powered cars would emit more CO_2 per vehicle-mile than Kelley (2020) had assumed based on figures published by AAA.

Both the DEIS and Kelley (2020) likely overestimated CO_2 emissions from cars in 2030 to 2045 for two reasons. Both ignored that gas-powered cars may be more fuel-efficient in the future and that electric cars may replace many gas-powered cars by 2030 or 2045. The problem with overestimating CO_2 emissions from cars is that it leads to overestimating the maglev's ability to reduce CO_2 emissions by reducing car travel.[11]

Net Increase in CO_2 Emissions from Maglev Operation

The last step is to sum the two CO_2 emission estimates just described in Appendix D4. Taken together, the slight emission reduction from reduced car travel and the larger emission increase from generating electricity to run the maglev would result in a net emission increase of 286 to 336 million kilograms of CO_2 per year ([286,336] = 460 - [124,174]).

The DEIS states that carbon dioxide is by far the dominant greenhouse gas related to maglev operation and so carbon dioxide is the only greenhouse gas that the DEIS estimates. The present author agrees that this simplification is reasonable.[12]

Avoid Becoming Confused by Table D4-44

As discussed thus far, Appendix D4 reports that an increase in greenhouse-gas emissions would result from maglev operation. The reader, however, could become confused by Table D4-44. Keep in mind that Table D4-44 does not show net greenhouse gas emissions. Instead, Table D4-44 is trying to show net air-pollution emissions that include both greenhouse and non-greenhouse gases.

Appendix D4 claims that Table D4-44 shows net savings from all air pollutants. It is clear from comparing this table to two other tables, however, that only one row includes all pollutants (CO, NO, PM2.5, VOC, and CO_2), specifically the auto-and-bus row. The maglev row of Table D4-44 contains only CO_2. The mismatch between the these two rows means that the table's Total row is meaningless. You can't compare apples and oranges. The two other tables that Table D4-44 is based on are Tables D4-43 and D4-40.[13]

In partial defense of Appendix D4, Chapter 4 did state that the "FRA did not quantify the power-plant emissions required for [maglev] train operations and facilities" (pg. 4.16-3). But, for this reason, the text of Appendix D4 has no business claiming that Table D4-44 shows net air-pollution emissions from operating the maglev. It makes no sense to title Table D4-44 "net emissions" because that title might confuse a casual reader into thinking that Table D-44 demonstrates that maglev operation would decrease net CO_2

---

11. 138 million kg y^{-1} = \$138,000 y^{-1} · \$1 t^{-1} · 1000 kg t^{-1}, where metric ton is abbreviated "t." \$1 per metric ton conversion factor: Appendix D4, Table D4-43, pg. D-52. \$138,000: Appendix D4, Table D4-40, pg. D-51. Kelley (2020) estimated only 59 million kilograms of car-related emission reduction rather than the DEIS's 138 million kilograms. In 2021, the DEIS estimated that the maglev would divert 316.1 million car vehicle-miles per year: Appendix D2, Table D2-3, pg. A-3. In 2015, BWRR estimated that the maglev would divert 165 million car vehicle-miles per year: Rogers 2015, pg. 11, 18.

12. CO_2 the only GHG modeled: Chapter 4.16, pg. 4.16-2.

13. CO is carbon monoxide, NO is nitric oxide, PM2.5 is fine particulate matter less than 2.5 micrometers in diameter, VOC stand for "volatile organic compounds," and CO_2 is carbon dioxide.

emissions. In contrast, Tables D4-40 and D4-43 establish that maglev operations would actually increase CO_2 emissions.

Appendix 2

CO_2 from Constructing the Maglev

This appendix estimates the CO_2 that would be emitted to construct a maglev track between Baltimore and Washington. A lower bound is calculated by estimating the CO_2 emissions that would result from manufacturing the steel and concrete to build just the maglev track. A higher, more comprehensive estimate is also provided, using the concept of emission intensity.

Lower Bound

A lower bound can be estimated for the CO_2 emissions from maglev construction by estimating the CO_2 that would be emitted only by manufacturing the concrete and steel needed to construct the tunnel and elevated track.

Two methods are used to give a lower bound of either 249 or 721 million kilograms of CO_2 released to construct the track. Both methods are similar in that they use the following equation to estimate the CO_2 emissions $m_{construct}$ (kg):

(1) *Emissions from constructing tunnel and elevated track*

$$m_{construct} = d\,(f_{tunnel}\,m_{tunnel} + f_{elevate}\,m_{elevate})$$

In this equation, d is the total length of the maglev track and f_{tunnel} and $f_{elevate}$ are the fractions of that length covered by an underground and elevated track, respectively. The terms m_{tunnel} and $m_{elevate}$ are the per-kilometer CO_2 emission rates for the tunnel and elevated track, respectively.

The methods differ in that they use a different estimate for m_{tunnel} and $m_{elevate}$. Method 1 derives values for m_{tunnel} and $m_{elevate}$ from the quantities in Table 2 and the following equations:

(2) *Emissions from concrete and steel used in construction*

$$m_{tunnel} = e_{concrete}\,r_{concrete,tunnel} + e_{steel}\,r_{steel,tunnel}$$
$$m_{elevate} = e_{concrete}\,r_{concrete,elevate} + e_{steel}\,r_{steel,elevate}$$

In method 1, Equations (2) and (1) are evaluated as follows:

$$13.08\times10^6\,\mathrm{kg\,km^{-1}} = 300\,\mathrm{kg\,m^{-3}} \cdot 30{,}000\,\mathrm{m^3\,km^{-1}}$$
$$+ 1.7\,\mathrm{kg\,kg^{-1}} \cdot 2.4\times10^6\,\mathrm{kg\,km^{-1}}$$

$$11.44\times10^6\,\mathrm{kg\,km^{-1}} = 300\,\mathrm{kg\,m^{-3}} \cdot 20{,}000\,\mathrm{m^3\,km^{-1}}$$
$$+ 1.7\,\mathrm{kg\,kg^{-1}} \cdot 3.2\times10^6\,\mathrm{kg\,km^{-1}}$$

$$721\times10^6\,\mathrm{kg} = 60.84\,\mathrm{km}\,(0.75 \cdot 13.08\times10^6\,\mathrm{kg\,km^{-1}}$$
$$+ 0.25 \cdot 11.44\times10^6\,\mathrm{kg\,km^{-1}})$$

Method 2 uses the values for m_{tunnel} and $m_{elevate}$ published in Kato and Shibahara (2005). Using their values, Equation (1) is evaluated as follows, and Equation (2) is not relevant:

$$249\times10^6\,\mathrm{kg} = 60.84\,\mathrm{km}\,(0.75 \cdot 5.31\times10^6\,\mathrm{kg\,km^{-1}}$$
$$+ 0.25 \cdot 3.68\times10^6\,\mathrm{kg\,km^{-1}})$$

Emission Intensity

The lower bound estimated in the preceding section is assuredly an underestimate of the emissions from constructing the maglev because it estimates emissions from manufacturing only the steel and concrete used and because it concerns itself with only the tunnel and elevated track, rather than all maglev-related infrastructure.

A more comprehensive but still rough estimate of 10.2 billion kilograms of CO_2 may be calculated using the concept of emission intensity. To calculate emission intensity, economists take

4 • GREENHOUSE GAS

Table 2: Parameters used in Appendix 2 to estimate the carbon-dioxide emission to construct the track of the proposed Baltimore-Washington maglev

Variable	Quantity	Notes
d	60.84 km	The maglev track that would connect Baltimore & Washington would be 33–36-mile long according to the draft environmental impact statement (FRA 2021, Chapter 3, pg. 3-18 and 3-19). Convert 36 miles to kilometers by multiplying by 1.609.
f_{tunnel}	0.75	The fraction of the track that would be in a tunnel or along an elevated track. The route would be 67% to 80% in a tunnel and the rest would be elevated (FRA 2018, pg. 9,12).
$f_{elevate}$	$1 - f_{tunnel}$	
Method 1: m_{tunnel} and $m_{elevate}$ derived from the concrete and steel to build the track		
$e_{concrete}$	300 kg m^{-3}	Fantilli et al. (2019) and Gursel (2014, Fig. 5.32) report 200 to 400 kg of CO_2 are emitted to manufacture 1 m^3 of concrete.
e_{steel}	1.7 kg kg^{-1}	Fantilli el al. (2019) and Liu et al. (2019) report 1.38 to 2.0 kg of CO_2 are emitted to a 1 kg of steel.
$r_{concrete,tunnel}$	30,000 m^3 km^{-1}	The amount of concrete or steel required to build 1 km of tunnel or elevated track (IEA 2019, pg. 57).
$r_{concrete,elevate}$	20,000 m^3 km^{-1}	
$r_{steel,tunnel}$	2.4×10^6 kg km^{-1}	
$r_{steel,elevate}$	3.2×10^6 kg km^{-1}	
Method 2: published values used for m_{tunnel} and $m_{elevate}$		
m_{tunnel}	5.31×10^6 kg km^{-1}	The amount of CO_2 that would be emitted to build 1 km of tunnel or elevated track (Kato and Shibahara 2005).
$m_{elevate}$	3.68×10^6 kg km^{-1}	

annual CO_2 emissions in a sector of the economy and divide it by the dollars spent in that sector. For the US construction industry, the emission intensity is in the vicinity of 0.6 kilograms of CO_2 per dollar of economic activity (0.6 kg/$: Huang et al. 2018, Fig. 2; 0.37-0.49 kg/$: EPA 2009, Table 3 and Fig. 4).

If CO_2 emissions from constructing the maglev occurred at this industry-wide rate, the maglev's $17-billion price tag would result in 10.2 billion kilograms of CO_2 emission (10.2 = 17 · 0.6). This amount of emissions would be more than a factor of ten greater than the lower bound estimated in the previous section.

The 10.2-billion-kilogram estimate is a rough estimate for emissions from all aspects of constructing a maglev between Baltimore and Washington, including all activities that the $17 billion would be spent on. This set of tasks would include construction planning and management; manufacturing the train; building the track, tunnel, stations, control facility, and maintenance faculties; and the fuel used by on-site construction machinery.

Furthermore, if actual construction were to cost $25.5 billion instead of $17 billion, i.e., 50% more than the estimate in the draft environmental impact statement, then the CO_2 emissions estimate would increase proportionally to 15.3 billion kilograms.

Cost overruns of this magnitude are not uncommon in the transportation industry. A number of studies found that construction costs for rail projects are, on average, about 50% higher than estimated prior to construction. Various theories have been proposed to explain why cost estimates continue to have a low bias in environmental impact statements even after researchers have published evidence that the bias exists.[14]

Projects similar to the proposed Baltimore-Washington maglev tend to have higher-than-average cost overruns, more than double the initial estimate in one case. Before construction began, the California high-speed rail from Anaheim to San Francisco was expected to cost $33.6 billion in 2008. Today, the estimate has more than doubled to $80 billion with the track partially built. The maglev planned in Munich was initially expected to cost 1.85 billion euros but it was canceled in 2008 as construction was about to begin in part because the cost estimate had risen to 3 billion euros. The Tokyo-Osaka maglev was expected to cost 5.1 trillion yen in 2007, but estimates rose to 9.1 trillion yen as construction proceeded.[15]

Any emission estimate calculated using industry-average emission intensity is only a rough approximation because a dollar spent on the construction of maglev-related infrastructure might have a much higher or lower emission intensity than the US construction industry overall.

A few additional aspects of emission intensity are worth keeping in mind. When researching emission intensity, it is important to keep track of units. Sometimes emission intensity is expressed as the mass of CO_2 emitted per dollar spent; other times, it is expressed as the mass of CO_2 emitted

---

14. 45% cost overrun for rail projects: Priemus et al. (2008). 50% cost overrun for rail projects: http://americandreamcoalition.org/?page_id=3813. Persistence of low bias in EIS cost estimates: Sturm et al. (2011).

15. California high-speed rail estimate of $33 then $80 billion: Editorial, Los Angeles Times, 17 June 2020. Munich Link Transrapid maglev cost overrun: https://en.wikipedia.org/wiki/Transrapid. The Tokyo-Osaka line is a superconducting maglev and is called the Chuo Shinkansen. Wikipedia.org states the cost estimate of 5.1 to 9.1 trillion yen.

per megajoule of electricity used in a process or project. Emission intensity expressed in metric tons per $1,000 is equivalent to emission intensity expressed in kilograms per dollar.

The emission intensity of the US construction industry is about twice that of the US economy overall. Approximately 0.27 kg of CO_2 is emitted per dollar of US gross domestic product (GDP), according to the World Bank (2016). The construction industry's 0.6-kg-per-dollar emission is composed of 0.33 kg per dollar of direct emissions and 0.27 kg per dollar of indirect emissions, according to Huang et al. (2018, Fig. 2). Direct emissions occur at the construction site. Indirect emissions occur elsewhere, including emissions to manufacture construction materials and generate the electric power used by construction-related activities.

On the question of highway congestion relief, many studies estimate that HSR [high speed rail] will have little positive effect because most highway traffic is local and the diversion of intercity trips from highway to rail will be small.

— Congressional Research Service (2009, pg. 14)

5 • ROAD CONGESTION

Data from the Federal Railroad Administration shows that building a maglev would do little to reduce regional road congestion

The draft environmental impact statement forecasts that the proposed maglev would have an insignificant impact on road congestion in the Baltimore-Washington region

---

Road congestion would not improve significantly if a superconducting magnetic-levitation rail line were built between Baltimore and Washington. A federal regulatory agency reviewed this "maglev" proposal and found the road congestion impact from building the maglev would be insignificant. The catch is that the review's executive summary states there would be an improvement in road congestion but buries in its appendices the evidence that the improvement would be so small that it would be barely noticeable.

Sloppy thinking on this point is harmful. Some elected officials and community leaders may have been persuaded to endorse the maglev because of the project's supposed traffic-reduction ability. A lot of money is at stake—$15 to $17 billion to build the maglev—and the Baltimore-Washington region does have a road congestion problem.

The regulatory agency in question is the Federal Railroad Administration. In January 2021, it published a draft environmental impact statement for the proposed Baltimore-Washington maglev. The present chapter examines the impact statement's road-congestion forecasts.

This chapter provides historical context by noting several studies that found that constructing high-speed rail would be unlikely to reduce road congestion significantly. Also relevant is the rather small road-congestion impact forecast in 2015 by Baltimore Washington Rapid Rail, the company that wants to build this maglev.

Historical Context

There is long-standing doubt about whether high-speed rail (HSR), in general, can significantly reduce road congestion. High-speed rail includes both maglevs and conventional steel-wheel trains. In 2009, the Congressional Research Service wrote:

> On the question of highway congestion relief, many studies estimate that HSR will have little positive effect because most highway traffic is local and the diversion of intercity trips from highway to rail will be small. (pg. 14)

Years earlier, the Federal Railroad Administration gave a brief explanation of why fast trains are unlikely to reduce roadway traffic by a significant amount:

> The ability to divert patrons from existing modes depends not on line-haul times, but on comparative total travel times, which also include access to, egress from, and time spent in stations. The composition of those total travel times varies dramatically among modes.... In any comparison of total timings, auto has an inherent advantage in its door-to-door convenience (avoiding access and terminal time) (FRA 1997, pg. 7-4)

The Federal Railroad Administration's 2021 draft environment impact statement for the Baltimore-Washington maglev makes no mention of the above-cited studies, but an FRA report in 2008 did elaborate on this point:

> Automobile travel differs from air or rail travel in that it generally involves door-to-door service, offers greater flexibility in time of departure, and does not require travelers to share space with strangers. Consequently, rail travel must be extremely competitive in other dimensions, such as speed or cost, to attract automobile travelers. (FRA 2008, pg. 6-7)

With this context in mind, we turn to data specific to the proposed Baltimore-Washington maglev.

BWRR Testimony

In 2015, Baltimore Washington Rapid Rail (BWRR) testified before the Maryland Public Service Commission about the benefits that the maglev would provide to Baltimore and Washington. BWRR argued that the region's traffic was bad and getting worse, and the company implied that the maglev could contribute significantly to solving the congestion problem.

To support the idea that the maglev would significantly reduce the region's traffic woes, BWRR stated that the region's car traffic would increase by 34% over the next 25 years were the maglev not built. BWRR also stated that operating the maglev would reduce the number of miles that cars were driven in the region by 165 million vehicle-miles per year relative to the no-maglev scenario (Rogers 2015, pg. 11 and 18). To be clear about the units being used here, the quoted figure is in vehicle-miles, which are calculated by counting each mile that a car moves in contrast to passenger-miles, which are calculated by multiplying the number of miles that a car travels by the number of people in it.

This traffic reduction sounds significant until you do the math, but BWRR's testimony left out one data point that is required to do the math. This data point is available elsewhere: cars are currently driven approximately 44.4 billion vehicle-miles per year in the Baltimore-Washington region (FRA 2021, Chapter 2, pg. 2-7).

Combining BWRR's numbers, we arrive at an estimate for the annual percent increase in regional car traffic without the maglev and the percentage offset in car traffic that would occur were the maglev to begin operation. BWRR's numbers give an annually compounded rate of increase in car traffic in the Baltimore-Washington region of 1.12% without the maglev ($1.0112 \approx 1.34^{(1 \div 25)}$). BWRR's numbers also give an offset of 0.37%

less regional car traffic were the maglev operating (0.37%=100% {0.165/44.4}).

Comparing 1.12% to 0.37%, reveals that a few months of the natural increase in the region's road traffic would erase the traffic-reduction benefit of the maglev (4 months ≈ 12 months · 0.37% ÷ 1.12%). This result from the BWRR numbers is shown in the bottom row of Table 1.

In light of this comparison, BWRR has failed to make a convincing argument that the maglev would reduce road congestion to a meaningful degree. Why didn't journalists dig into BWRR's 2015 testimony years ago and reveal this fact?

Taking a step back, the reason why the maglev could reduce road congestion at all is that some existing car travelers are forecast to switch from car to maglev. In discussing this transportation choice, it is inaccurate to describe it as a choice between car or maglev travel. Instead, a more accurate description would be a choice between either driving directly to the destination or taking a "maglev assisted" trip. Using the maglev may involve car travel, including driving to reach the initial maglev station and taking a taxi or car service from the final maglev station to the destination.

Regional Impact

The maglev's inability to significantly reduce road congestion is established in the appendices of the draft environmental impact statement (DEIS) published by the Federal Railroad Administration in January 2021.

The best place to start may be the DEIS estimate of the maglev's impact on car traffic in the entire Baltimore-Washington region. The region's current car traffic is 44.4 billion vehicle-miles per year (44.4 = 25.2 + 19.2, Chapter 2, pg. 2-7). The implied change in car traffic during the next 25 years if the maglev were not built is approximately

Table 1. The traffic impact of the proposed Baltimore-Washington maglev as estimated over the Baltimore-Washington region, over the corridor that contains the maglev track, or at a point on Interstate 95 that lies between Baltimore and Washington.

Area or location	Annual car-traffic volume at start of period[a]	Car-traffic baseline annual increase without maglev	Car-traffic: maglev offset vs. baseline[b]	Duration of maglev-induced car-traffic reduction[c]
2021 DEIS				
1. Baltimore-Washington region	44.4 billion	+0.99%	-0.71%	9 months
2. Baltimore-Washington corridor	2.9 billion	+0.37%	+0.47%	no reduction
3. Interstate 95 at Route 100	79.9 million	+0.16%	-0.36%	2 years
2015 BWRR				
4. Baltimore-Washington region	--	+1.12%	-0.37%	4 months

[a] Units: vehicle-miles per year for rows 1, 2, and 4, and vehicles per year for row 3. The 2021 DEIS refers to car vehicle-miles as VMT or "vehicle miles traveled."

[b] Car traffic percent offset if the maglev were built relative to the no-maglev scenario.

[c] The duration is calculated by dividing the maglev percent offset by the baseline percent annual increase.

28±5%, given that the DEIS states that the region's population and employment will increase by 23% and 33%, respectively (28% = {23% + 33%} ÷ 2, Chapter 2, pg. 2-2). The DEIS states that maglev operation would reduce baseline car traffic by 316.1 million vehicle-miles per year (Appendix D2, Table D2-3, pg. A-3). The present chapter quotes the DEIS estimate for the Camden Yards station option.

Now let's transform this information into a form that makes comparisons easier. First, the regional road traffic would increase at an annually compounded rate of 0.99% ($1.0099 \approx 1.28^{(1 \div 25)}$). Second, if the maglev were operating, regional car traffic would be reduced by 0.71% below the no-maglev car-traffic baseline (0.71% = 100% {0.3161 ÷ 44.4}). The result is that, in about nine months, the natural increase in the region's car traffic would erase the congestion-reduction power of the maglev (9 months ≈ 12 months · 0.71% ÷ 0.99%). These DEIS results are shown in the first row of Table 1, which can be contrasted with the previously discussed BWRR values that are shown in the bottom row of Table 1. To visualize this small and short-lived reduction in road congestion, compare the dotted line and solid line in Figure 1.

If the maglev's ability to reduce regional road congestion were erased after nine months, the maglev would be an ineffective solution, given that construction would cost $15 to $17 billion. The exact cost is unclear. There appears to be a typographical error in the DEIS executive summary where it states a lower construction cost of $10–$13 billion. Even that cost, however, would be embarrassingly high for such a small impact on road congestion. It is unclear from the DEIS and media reports what fraction of the construction cost would be borne by the US taxpayer, private investors, or a foreign government seeking to promote its maglev technology.[1]

Comparing rows 1 and 4 of Table 1, the road-congestion reduction that was forecast in the 2021 DEIS is approximately double the forecasted impact in BWRR's 2015 testimony that was discussed earlier. Nonetheless, both predictions represent a small and ephemeral reduction in the region's road congestion. Both the BWRR and DEIS forecasts of road-congestion impact may be overly optimistic because the official maglev ridership forecast is likely too high by a factor of 10, as discussed in Chapter 1.

Corridor Impact

Along the roads within the corridor that contains maglev stations and track, car traffic is forecast to worsen if the maglev were built even though the maglev is forecast to improve road traffic slightly over a larger area, namely the entire Baltimore-Washington region. To reach this conclusion, the 2021 draft environmental impact statement (DEIS) examines the narrow corridor that includes the length of the maglev track and a swath of land extending at least a quarter mile away from the track, and in some places, extends several miles away. The DEIS defines the boundaries of this corridor and refers to it as the "mesoscale subarea" (Appendix D9, pg. D9-27).

In this corridor, current car traffic is 7.979 million vehicle-miles per day, forecast to increase to 8.530 million vehicle-miles per day over 18 years if the maglev were not built (Appendix D9, Table D9-9, pg. D9-50). This increase works out to an annually compounded rate of 0.37% (1.0037

---

1. $15–$17 billion: Appendix D4, Table D4-8, pg. D-21. $11–$13 billion: Executive Summary, Table ES43-2, pg. ES-20.

≈ 1.069 $^{(1\div18)}$ and 6.9% = 100% {8.530 − 7.979} ÷ 7.979).

If the maglev were built, the DEIS forecasts that car traffic within the corridor would actually be greater, not less, than in the baseline no-maglev scenario. Operating the maglev would generate additional car traffic in the corridor: 37.3 thousand vehicle-miles per day (Appendix D9, Table D9-9, pg. D9-50). These values are summarized in row 2 of Table 1 of the present chapter, with miles per day converted to miles per year by multiplying by 365.25 days per year.

The maglev would worsen road congestion in the corridor because the corridor contains the proposed maglev stations. Car traffic would worsen near a maglev station as people divert from driving directly to their various destinations and instead all converge on the station. The following section examines this kind of local impact.

Point Impact

The 2021 draft environmental impact statement (DEIS) states that the maglev would worsen road congestion in some locations and improve it slightly in others. The maglev would worsen traffic jams near maglev stations while reducing traffic on the stretch of Interstate 95 that most people use when driving between Baltimore and Washington. These impacts are described below in more detail.

Discussing the proposed downtown Washington maglev station at Mount Vernon Square, the 2021 DEIS clearly states that nearby traffic would worsen because of the maglev. The DEIS forecasts that customers traveling to and from this station by car, when added to existing road traffic, would cause motorists to experience intolerable delays near the station (Chapter 4.2,

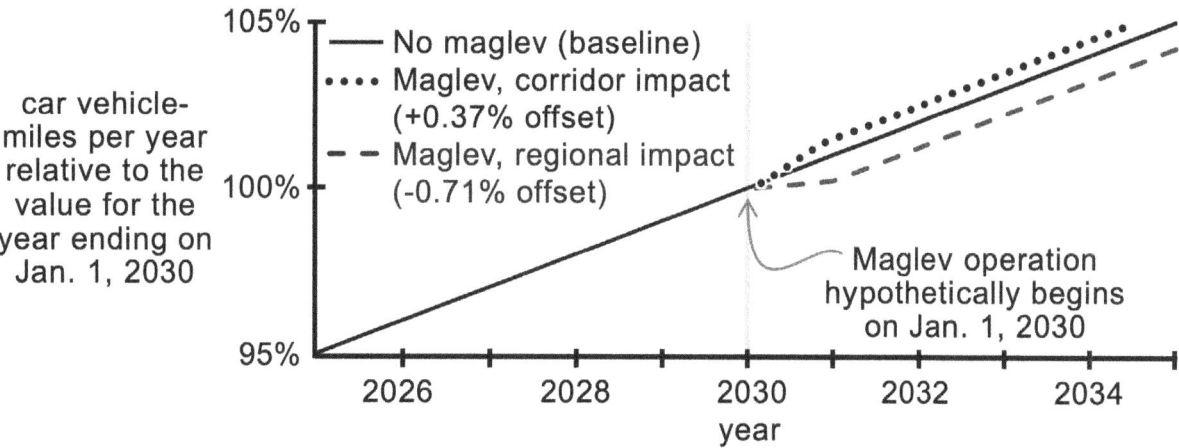

Figure 1. Schematic representation of the proposed Baltimore-Washington maglev's impact on road congestion based on data in the Federal Railroad Administration's 2021 draft environmental impact statement (DEIS). In a narrow corridor along the maglev route, the DEIS forecasts that the maglev would cause car traffic to increase (dotted line) relative to the baseline no-maglev scenario (solid line). In a wider region, the DEIS forecasts that the maglev would reduce car traffic (dashed line) relative to the baseline no-maglev scenario, but car traffic would still increase in absolute terms.

Table 4.2-6, pg. 4.2-25). A footnote directs us to interpret a grade of "E" or "F" at locations near the station as meaning that intolerable delays would occur were the maglev built. The congestion would become so severe in the vicinity of Mount Vernon Square that the DEIS recommends measures such as "encourage drivers through public outreach efforts to choose alternative routes in order to avoid the station area to the degree possible" (Chapter 4.2, pg. 4.2-26).

The DEIS text appears somewhat garbled regarding the traffic impacts near the proposed maglev station at Baltimore's Camden Yards. Appendix D2 states, "the data in the table shows significant deterioration in traffic operations" due to the maglev, but some of the entries in the cited table show only minor impact. One could speculate that this table and the associated text were in flux shortly before the DEIS was published. A footnote states, "additional coordination with the project sponsor is underway" (Appendix D2, Table D2-18, pg. A7-29).

The just-cited Appendix D2 data for maglev stations is reflected to varying degrees of accuracy in the body of the DEIS. In Chapter 4.2, one page states that the maglev's traffic impact would be "positive," while a few pages later, the text states that the impact would be "minimal" (Chapter 4.2, pg. 4.2-8 and 4.2-20).

In its evaluation of Interstate 95, the 2021 DEIS describes what may be close to the largest-possible car-traffic reduction that the maglev could produce. Specifically, the DEIS describes the traffic impact at a point along Interstate 95 near Elkridge, a location through which most car travel between Baltimore and Washington passes.

The DEIS finds a rather small reduction in car traffic on Interstate 95 at Route 100 in Elkridge. At this location, current traffic is 218,700 cars per day (Appendix D2, Table D2-16, pg. A5-26). The DEIS forecasts that cars per day would increase by 0.16% each year at this location were the maglev not built. The 0.16%-per-year increase comes from the DEIS-supplied 4.07% increase in 25 years ($1.0016\% = 1.0407^{(1\div25)}$ and $4.07\% = 100\% \{227.6 - 218.7\} \div 218.7$). The DEIS explicitly states that operating the maglev would reduce car traffic below the baseline level by an offset of 0.36% (Appendix D2, Table D2-16, pg. A5-26). These values are summarized in row 3 of Table 1 of the present chapter.

The DEIS forecast for this point on Interstate 95 means that two years after the maglev begins operation, car traffic would be back where it was when the maglev started operating. After that, car traffic would be worse than before. Two years is a brief respite from worsening traffic considering the great cost of building the maglev.

Implications of these Results

The present chapter examines road-congestion data in the 2021 draft environmental impact statement (DEIS) and also in the 2015 testimony of Baltimore Washington Rapid Rail. The data in these documents establish an insignificant and fleeting reduction in regional road congestion if the proposed Baltimore-Washington maglev were built and operated.

If this finding became widely known, it would hamper maglev marketing efforts because, if the prospect of reducing road congestion were removed, negative aspects of the project would make a bigger impression. Negative aspects include a construction cost of $15 to $17 billion and a one-way ticket price of $40 to $80 per person.[2]

---

2. Construction cost: Appendix D4, Table D4-8, pg. D-21. Ticket price: Appendix D2, pg. D-108.

In light of this finding, it seems odd that the executive summary of the 2021 DEIS failed to mention two things. First, the executive summary failed to mention the relevant data in the appendices that establish that the maglev would do little to improve regional road congestion.

Second, the DEIS executive summary failed to mention that the DEIS Project Need section emphasizes transportation delays as the central need that the DEIS is intended to addressed. The Project Need section states that action is needed to address the following transportation issues and challenges:

- The Baltimore-Washington region makes up one of the largest and densest population centers in the United States
- Travel demand will continue to increase
- Inadequate capacity of the existing transportation network (Chapter 2, pg. 2-2)

The Federal Railroad Administration would find it difficult to delete the above-quoted text because the DEIS is tied to the notice of intent published in the Federal Register (25 Nov 2016, pg. 85, 320). The notice of intent has its own Purpose and Need section that states, "demand on the transportation infrastructure between Baltimore and Washington will continue to increase... thereby decreasing the level of service.... As congestion increases... continued economic development will be impacted."

The word "congestion" is used only twice in the DEIS executive summary. The executive summary states that the DEIS focuses on the Baltimore-Washington region because transportation congestion here is severe, a point on which everyone agrees. The executive summary also states that the maglev would enable people to "bypass congested locations." This statement is a weak substitute for a quantitative forecast of the maglev's impact on regional road congestion—especially considering that the executive summary could have quoted the relevant forecasts found in the DEIS appendices.

The DEIS executive summary would more effectively communicate road congestion impacts if it included a schematic diagram. Specifically, a diagram similar to Figure 1 of the present chapter. Figure 1 depicts how the maglev would not reduce regional road congestion in absolute terms and how the maglev would do little to slow the rate at which regional road congestion continues to worsen.

Figure 1 depicts a hypothetical 1% annual increase in car traffic if the maglev were not built (solid line). This percentage increase is similar to the baseline car-traffic growth rates listed in Table 1 (page 85). In Figure 1, the dashed line represents the DEIS forecast for the regional impact of the maglev: an offset of -0.71% from the baseline no-maglev scenario. The dotted line depicts the DEIS forecast for the impact of the maglev along the corridor that includes the maglev stations and track and their immediate neighborhood. As discussed earlier in the present chapter, the DEIS-estimated corridor impact is an offset of +0.37% from the baseline no-maglev scenario.

Within reason, the exact percentages used in Figure 1 are irrelevant because the conclusion would remain the same. After a few years—or perhaps only a few months—of operating the proposed maglev, car traffic would become worse than before the maglev was built.

Conclusion

Based on the 2021 draft environmental impact statement (DEIS), it is clear the proposed Baltimore-Washington superconducting maglev would do little to improve regional road congestion. By altering traffic patterns without significantly reducing the number of cars on the road, the maglev is likely to worsen road congestion in some places, such as near the maglev stations in downtown Washington and Baltimore. The relevant data are published in the DEIS appendices, but the DEIS executive summary does not reflect these data.

The DEIS correctly states that traffic jams reduce the quality of life in the Baltimore-Washington region, but it is odd that the DEIS focuses on one, somewhat eccentric proposal to address this problem and entirely ignores other, more common-sense approaches.

The 2021 DEIS fails to provide decision-makers and the public a comparison of the proposed maglev's effectiveness at solving the region's congestion problem relative to the effectiveness of other solutions. The DEIS provides no evidence that the maglev would be more cost effective than, for example, promoting teleworking, modifying the interchanges that produce the worst rush-hour bottlenecks, or funding projects that could help create affordable and desirable neighborhoods close to employment centers.

References

Congressional Research Service, 2009: *High Speed rail (HRS) in the United States.* D. Randall, J. Frittelli, and W. J. Mallett, Report to Congress, 7-5700, R40973, https://fas.org/sgp/crs/misc/R40973.pdf.

Federal Railroad Administration, September 1993: *Final Report on the National Maglev Initiative (NMI).* Technical Report DOT/FRA/NMI-93/03. 121 pp, https://railroads.dot.gov/elibrary/final-report-national-maglev-initiative.

Federal Railroad Administration, January 2021: *Draft Environmental Impact Statement and Draft Section 4(f) Evaluation, Baltimore-Washington Superconducting Maglev Project.* https://bwmaglev.info/index.php/project-documents/deis.

Federal Railroad Administration, 2008: *Analysis of the Benefits of High-Speed Rail on the Northeast Corridor.* Report CC-2008-091, memorandum from D. Tornquist, 19 pp, https://www.oig.dot.gov/library-item/30401.

Federal Railroad Administration, 1997: *High-speed Ground Transportation for America.* 182 pp, https://railroads.dot.gov/sites/fra.dot.gov/files/fra_net/1177/cfs0997all2.pdf.

Federal Register, 25 November 2016: *Environmental Impact Statement for the Baltimore-Washington Superconducting Maglev (SCMAGLEV) Project, Between Baltimore, Maryland, and Washington, DC.* pp. 85319–85321, https://bwmaglev.info/index.php/project-documents/reports.

Rogers, W., 17 April 2015: Direct testimony of Wayne L. Rogers, Case #9363. Maryland Public Service Commission, 23 pp, https://www.psc.state.md.us/search-results/?q=9363&x.x=20&x.y=20&search=all&search=case.

ABOUT THE AUTHOR
Owen A. Kelley

Education
- Ph.D. Computational Sciences and Informatics, George Mason University, Fairfax, Virginia (2008).
- M.S. Applied Physics, George Mason University (1997).
- B.A., St. John's College, the "Great Books" school in Annapolis, Maryland (1993).

Professional Experience
- Assistant Research Professor, George Mason University (1997–present). This position is located off campus at the NASA Goddard Space Flight Center in Greenbelt, Maryland.
- Research Assistant, Upper Atmospheric Physics Branch, Naval Research Laboratory (1993–1996), Washington, DC.

Selected Publications
- Kelley, O. A., 2014: Where the least rainfall occurs in the Sahara Desert, the TRMM radar reveals a different pattern of rainfall each season. *Journal of Climate*, **27**, 6919–6939.
- Kelley, O. A., and J. B. Halverson, 2011: How much tropical cyclone intensification can result from the energy released inside of a convective burst? *Journal of Geophysical Research*, **116**, D20118, doi:10.1029/2011JD015954.
- Kelley, O. A., J. Stout, M. Summers, and E. J. Zipser, 2010: Do the tallest convective cells over the tropical ocean have slow updrafts? *Monthly Weather Review*, **138**, doi:10.1175/2009MWR3030.1.
- Kelley, O. A., J. Stout, and J. B. Halverson, 2005: Hurricane intensification detected by continuously monitoring tall precipitation in the eyewall. *Geophysical Research Letters*, **32**, L20819, doi:10.1029/2005GL023583.

Selected Media Interactions
- *Capital Connection's Asia* program on CNBC Asia (11 September 2017). Interviewed by anchor Nancy Hungerford about NASA observations of Hurricane Irma.
- *Hurricane: The Anatomy* (2014). Interviewed for a hurricane documentary directed by Andy Byatt of Saint Thomas Productions.
- *Science Friday* (2 November 2012). Interviewed for a companion video to the *Science Friday* radio program, discussing Hurricane Sandy observations made one day before it struck New York City.
- "Hyperhurricane" (2007). Interviewed for this episode of the *Naked Science* series aired on the National Geographic Channel.

Volunteer Activities
- Self-published a 246-page field guide, *A Hundred Wild Things: A Field Guide to the Plants of the Greenbelt North Woods* (2019).
- Member, Greenbelt Forest Preserve Advisory Board (June 2017–present).
- Analyzed various aspects of the proposed Baltimore-Washington superconducting maglev. Wrote articles and letters to the editor in the *Greenbelt Online* blog, Prince George's County Sierra Club Issues Forum, *Capital Gazette*, and *Greenbelt News Review* (2017–present).
- Wrote ecology-related articles published in the *Greenbelt News Review* (2017–2020).

www.ingramcontent.com/pod-product-compliance
Lightning Source LLC
Chambersburg PA
CBHW061821290426
44110CB00027B/2939